# The Cold War Politics of Genetic Research

# *Archimedes*

NEW STUDIES IN THE HISTORY AND PHILOSOPHY OF SCIENCE AND TECHNOLOGY

VOLUME 24



*Archimedes* has three fundamental goals; to further the integration of the histories of science and technology with one another: to investigate the technical, social and practical histories of specific developments in science and technology; and finally, where possible and desirable, to bring the histories of science and technology into closer contact with the philosophy of science. To these ends, each volume will have its own theme and title and will be planned by one or more members of the Advisory Board in consultation with the editor. Although the volumes have specific themes, the series itself will not be limited to one or even to a few particular areas. Its subjects include any of the sciences, ranging from biology through physics, all aspects of technology, broadly construed, as well as historically-engaged philosophy of science or technology. Taken as a whole, *Archimedes* will be of interest to historians, philosophers, and scientists, as well as to those in business and industry who seek to understand how science and industry have come to be so strongly linked.

For further volumes:
http://www.springer.com/series/5644

William deJong-Lambert

# The Cold War Politics of Genetic Research

## An Introduction to the Lysenko Affair

Springer

William deJong-Lambert
Bronx Community College, CUNY
History
University Avenue, NY
USA

ISSN 1385-0180
ISBN 978-94-007-2839-4 e-ISBN 978-94-007-2840-0
DOI 10.1007/978-94-007-2840-0
Springer Dordrecht Heidelberg London New York

Library of Congress Control Number: 2011945850



Printed on acid-free paper

Springer is part of Springer Science+Business Media (www.springer.com)

*To the tooth fairy*

# Acknowledgements

This book turned out, fortunately, nothing like I had intended. For this I have a few individuals to thank. First and foremost is Jed Z. Buchwald, whose support, persistence and patience were essential to my finally producing this manuscript. If I someday do someone else the same favor, I will have done a lot. Second is the anonymous peer reviewer who would not take "yes" for an answer and also pushed me to make this book into something I had never imagined. I must also thank Nikolai Krementsov, whose support in this and all of my other endeavors has been inspirational and essential, as well as Miklos Muller, whose company and conversation over countless lunches at Rockefeller University and sushi dinners at Totya, on the corner of 63rd and Third, have been essential to understanding what "Lysenkoism" was.

The help of several organizations and archivists was also essential to the success of this project. At the American Philosophical Society, where I began research for this book as a Library Resident Fellow nine years ago, I would like to thank Martin Levitt, Valerie-Ann Lutz, Michael Miller and Earle Spamer. A Fulbright-Hays Doctoral Dissertation Research Grant then supported a year of research in Warsaw, where I examined the impact of Lysenko's anti-genetics campaign upon Polish biology. A Franklin Research Grant from the American Philosophical Society enabled me to conduct further work in the Julian Huxley papers at the Woodson Research Center in the Fondren Library at Rice University and the J.B.S. Haldane papers at University College London. At the Woodson, I would like to thank Amanda Focke, Lauren Meyers and Dara Flinn. In Special Collections at University College London I would like to thank Deborah Furness, Dan Mitchell and Mandy Wise. Research in the Hermann J. Muller papers at the Lilly Library at Indiana University was supported by an Everett Helm Visiting Fellowship, and I would like to thank Zachary Downey, Sue Presnell and Cherry Dunham Williams for their continued assistance.

If had to cite a single source that has most influenced my thinking on the Lysenko controversy it would be the International Workshop on Lysenkoism, held December 4-5, 2009, at the Graduate Center of the City University of New York and the Harriman Institute at Columbia University. This meeting would not have been possible if not for the support of Catherine Nepomnyashchy at the Harriman and Gillian

Small, Vice Chancellor for Research at the City University of New York. I would also like to thank former-Bronx Community College President, Carolyn Williams, as well as Joe Dauben and Brian Schwartz at the Graduate Center, for their encouragement and assistance. I am also extraordinarily grateful to all of those who came to present their research, served as commentators or panel speakers, and contributed their comments and insights at the workshop. Among those in the latter category was my graduate school advisor, Gita Steiner-Khamsi, whose continued enthusiasm for my work has been extremely gratifying.

For the idea of writing a textbook on the Lysenko controversy I have two people to thank—Leslie Clarence Dunn and my wife Cheryl. In his oral history interview Dunn talked about the impact that writing one of the first genetics textbooks had upon the development of the field. The idea that a textbook, i.e., making something easier to teach, could play such an important role in the construction of knowledge, would never have occurred to me otherwise. That said, I would not have thought of this in terms of my own work had it not been for the numerous conversations Cheryl and I had about textbooks in which she informed me (as a former textbook editor) about how they are put together and the role they play in academia. Considering the fact that the next two people who come to mind as requiring acknowledgement are my children—Halina and Riley—who are always willing to give me the extra few seconds to finish a sentence before heading outside to hit the playground, makes me realize how much my family has contributed to my scholarship. I could not have done it without them.

# Contents

# Timeline of Events

1809: Jean Baptiste Lamarck publishes *Philosophie Zoologique*, outlining his theory of evolution based upon the inheritance of acquired characteristics

February 21, 1848: publication of Karl Marx and Friedrich Engels' *The Communist Manifesto*

November 24, 1859: publication of Charles Darwin's *On the Origin of Species by Means of Natural Selection, or the Preservation of Favoured Races in the Struggle for Life*

1866: Gregor Mendel's paper "Experiments on Plant Hybridization" published in *Proceedings of the Natural History Society of Brünn*

1869: Francis Galton publishes *Hereditary Genius*, founding the eugenics movement

1878: Publication of Friedrich Engels' *Anti-Dühring*

June 22, 1887: Julian Huxley born

December 21, 1890: Hermann Joseph Muller born

November 5, 1892: John Burdon Sanderson Haldane born

November 2, 1893: Leslie Clarence Dunn born

October 28, 1895: Conway Zirkle born

September 29, 1898: Trofim Denisovich Lysenko born

1900: Rediscovery of Mendel's Laws

January 25, 1900: Theodosius Dobzhansky born

1910: Eugenics Record Office founded at Cold Spring Harbor, Long Island

1912: Julian Huxley becomes chair of Biology at the newly-founded Rice Institute in Houston, Texas

1915–1918: H.J. Muller joins Huxley at the Rice Institute

November 7, 1917: Soviet power established in Russia

1918–1920: H.J. Muller returns briefly to teach at Columbia University then moves to the University of Texas at Austin the where he stays until 1932

1922: H.J. Muller brings *Drosophila* stocks to the Soviet Union

January 21, 1924: Vladimir I. Lenin dies

May 26, 1924: 1924 Immigration Act Passed by U.S. Congress limiting immigration to the United States from Southern and Eastern Europe, and Asia

July 10–21, 1925: Scopes Trial in Dayton, Tennessee

May 2, 1927: Supreme Court rules in Buck v. Bell that the Virginia statute authorizing sterilization does not violate the Constitution

August 7, 1927: First mention of Lysenko in *Pravda*

September, 1927: L.C. Dunn visits the Soviet Union for the first time

December 27, 1927: Theodosius Dobzhansky arrives in New York on a Rockefeller grant

1928: J.B.S. Haldane visits the Soviet Union at the invitation of Nikolai Vavilov; Theodosius Dobzhansky accompanies T.H. Morgan West to Cal Tech; L.C. Dunn joins the faculty at Columbia

October 24, 1929: Stock Market Crash in the United States, start of the Great Depression

November 7, 1929: Joseph Stalin's article "The Year of the Great Break" appears in *Pravda*

1930: Conway Zirkle accepts position as associate professor of botany at University of Pennsylvania

Summer, 1931: Julian Huxley visits the Soviet Union

1932: J.B.S. Haldane moves from Cambridge University to University College, London

August 24–31, 1932: VI International Congress of Genetics held in Ithaca, New York

September 5, 1932: H.J. Muller leaves the United States for Europe

September 16, 1933: H.J. Muller moves to the Soviet Union

1936–1938: The Great Terror

July 17, 1936–April 1, 1939: The Spanish Civil War

November, 1936: VII International Congress of Genetics to be held in Moscow cancelled

December 19–26, 1936: H.J. Muller participates in debate with T.D. Lysenko at Lenin All-Union Academy of Agricultural Sciences

March–May, 1936: H.J. Muller joins International Brigade in Spain, briefly visits the United States and returns to the Soviet Union in September

September 23, 1937: H.J. Muller leaves the Soviet Union for good

November, 1937–August, 1940: H.J. Muller works at the University of Edinburgh, Scotland

August 15–29, 1939: VII International Congress of Genetics held in Edinburgh, Scotland

September 1, 1939: Nazi Germany invades Poland, World War II begins

October 7–14, 1939: Nikolai Vavilov and T.D. Lysenko participate in discussion on "issues in genetics," sponsored by *Under the Banner of Marxism*

1940: Dobzhansky joins Dunn at Columbia

August 14, 1945: Japan surrenders after atomic bombs are dropped on Hiroshima and Nagasaki, World War II ends

September, 1945: H.J. Muller takes position at Indiana University, Bloomington, where he remains for the rest of his career

October, 1945: Anton Zhebrak publishes article in *Science* discounting Lysenko's impact upon Soviet genetics

1946: *Heredity and Its Variability*, written by T.D. Lysenko, translated into English by Theodosius Dobzhansky, is published by King's Crown Press; Julian Huxley begins two-year term as director of the newly-founded United Nations Educational, Scientific and Cultural Organization (UNESCO)

December, 1946: H.J. Muller awarded the Nobel Prize

May 14, 1948: State of Israel founded

February, 1948: Communists take power in Czechoslovakia

June 24, 1948: Soviets blockade Berlin until May 5, 1949

July 7–14, 1948: VIII International Congress of Genetics held in Stockholm, Sweden

July 31–August 7, 1948: Lenin All-Union Academy of Agricultural Sciences in Moscow holds a meeting "On the Situation in Biological Science"

1950: Mendel Semi-Centennial held to celebrate 50 years since the rediscovery of Mendel's Laws

February 9, 1950: U.S. Senator Joseph McCarthy, speaking before the Republican Women's Club of Wheeling, West Virginia, waves a piece of paper he claims contains a list of communists working in the State Department

March 5, 1953: Stalin dies

April 25, 1953: James Watson and Francis Crick publish their account of the double-helix structure of DNA

May 17, 1954: United States Supreme Court rules in landmark civil rights case, Brown v. the Board of Education, that separate but equal is inherently unequal, ending the legal system of segregation by race

October 23–November 4, 1956: Hungarian Revolution

October 29, 1956: Israel invades the Gaza Strip, initiating Suez Crisis, lasting until March, 1957

July 25, 1957: J.B.S. Haldane immigrates to India

October 4, 1957: Soviet Union launches Sputnik

April 12, 1961: Yuri Gagarin becomes first human to travel into space

1962: L.C. Dunn retires from Columbia; Theodosius Dobzhansky moves to the Rockefeller Institute until his retirement in 1971

December 2, 1964: J.B.S. Haldane dies

1965: U.S. President Lyndon Johnson awards Dobzhansky the National Medal of Science for his work in population genetics and the study of evolution

February 4, 1965: T.D. Lysenko removed from position as director of the Institute of Genetics of the Academy of Sciences

April 5, 1967: H.J. Muller dies

March 28, 1972: Conway Zirkle dies

March 19, 1974: L.C. Dunn dies

February 14, 1975: Julian Huxley dies

December 18, 1975: Theodosius Dobzhansky dies

November 20, 1976: T.D. Lysenko dies

# Introduction

*To that great company of Russian geneticists*
*and cytologists, now dispersed and destroyed,*
*to those who lost their positions and are denied*
*the exercise of their profession,*
*to those who simply disappeared,*
*to those who died under mysterious circumstances,*
*to those who, to save their families, recanted:*
*this book is respectfully dedicated.*

—Preface to Conway Zirkle's *Death of a Science in Russia*, 1949

Most readers consulting the 1961 *Encyclopedia Britannica* on Lamarckism probably did not notice the entry had been heavily revised from previous editions. That it would be updated is not surprising. A new name, Conway Zirkle, now appeared in the writer credits, next to T.H. Morgan, who had previously been listed as the sole author. Individual scientists are bound to interpret the history of their disciplines quite differently, and new collaborators often bring new ideas. What is striking in this instance however, is the nature of the edits, what is not discussed, and one detail many readers were probably unaware of.

Morgan's entry had focused exclusively on the persistence of belief in Lamarckism, as well as the scientific evidence against it. After reviewing popular examples (the blacksmith's son who inherits stronger arms from his father's repeated use of heavy hammers; the musically-gifted child who has the hours of time their parent spent practicing to thank for their talent), Morgan described Lamarck's influence upon Darwin, and later attempts to prove the inheritance of acquired characters by Kammerer, Dürken, Pavlov, and McDougall. Morgan concluded with a paragraph attributing the endurance of the doctrine to the fact that social evolution occurs thanks to the transmission of information from one generation to the next—stories, folk tales, historical works like the one you are reading right now—and explained that therefore it is natural we would apply this to the inheritance of physical features.[1]

In the 1961 edition, a few new sentences appeared amidst the discussion on Darwin, and the Soviet Union was mentioned for the first time.

> In the Soviet Union, for example, where the inheritance of acquired characteristics is accepted and where it has an official standing, it is presented as a part of the Darwinian theory and is referred to generally as "creative Soviet Darwinism," distinct from the "reactionary Darwinism" of capitalist countries.[2]

This point was elaborated further on in text inserted between what had been the penultimate and final paragraphs of the entry as it had been published in the previous edition, three years before. The new section described how a "prolonged and bitter scientific controversy raged in the Soviet Union from 1936 to 1948," which resulted in Lamarckism being "revived rather violently by the Communist authorities." According to the "authors," Marx and Engels were "staunch Lamarckians" because they believed "this type of inheritance would guarantee the future improvement of the human race," Soviet biologists supported it because it put them in a "strong tactical position in the socialist competition for status," and the outcome of the 1948 conference at the Lenin All-Union Academy of Sciences (VASKhNIL) where genetics was banned was a "boon to the communist theoreticians."[3]

While most of Conway Zirkle's analysis reflected his superficial views of the controversy, what is most interesting about the entry is something most readers would probably not have recognized: Morgan was dead when the entry appeared. Once again, there is not necessarily anything unusual about this. Very often authors and editors names remain attached to updated versions of works they no longer have a hand in writing or editing. In this case though two things are important. One, T.H. Morgan's name and reputation vastly exceeded Zirkle's, and gave the definition far greater authority than it would have were Morgan's name removed; Two, Morgan, had he been alive, would never have agreed to have his name attached to the version containing Zirkle's revisions.

It is obvious Zirkle must have been conscious of the first point. What did he need Morgan's name on the entry for anyway? He had published two articles on the history of the inheritance of acquired characteristics in *The American Naturalist*, as well as an essay in the *Transactions of the American Philosophical Society*.[4] It would have been no problem at all for him to entirely revise the entry and make it his own work. That, however, would give it less credibility because Morgan—as a Nobel Prize winner—was far more famous than Zirkle. As for the second point, Morgan's biographer Garland Allen has described how Morgan loathed political

[1] "Lamarckism," *Encyclopedia Britannica* 13 (1958): 607–10.

[2] "Lamarckism," *Encyclopedia Britannica* 13 (1961): 607.

[3] Ibid, p. 609.

[4] Conway Zirkle, "The Inheritance of Acquired Characteristics and the Provisional Hypothesis of Pangenesis," *The American Naturalist* 69, no. 724 (1935): 417–45; Conway Zirkle, "Further Notes on Pangenesis and the Inheritance of Acquired Characters," *The American Naturalist* 70, no. 731 (1936): 529–46.

activism. Morgan believed scientists should remain apolitical, because involvement in social causes had no place in scientific practice. In this instance it seems clear that Zirkle used Morgan to do something—attack "Marxian biology"—that Morgan, were he alive, would have wanted no part of.[5]

This brings us back to the above-quoted preface from Zirkle's 1949 book, *Death of a Science in Russia*. "Russian geneticists and cytologists, now dispersed and destroyed… lost their positions and are denied the exercise of their profession…," he wrote. But how many of them did Zirkle even know? To what extent did he really care? Was it not only their suffering that mattered to Zirkle, because it was useful to him?

When Zirkle wrote of, "those who died under mysterious circumstances," he was referring above all to renowned Russian geneticist Nikolai Vavilov. But is it not true that Vavilov was important to Zirkle only because he was gone? If Vavilov were alive, well and increasing agricultural production in the USSR, Zirkle would have attacked him. The thanatology of Lysenkoism is an important topic, but the more pressing question is—what was the "Lysenko affair,"[6] and why did anyone want to get involved in it?[7]

Zirkle, and *Death of a Science in Russia*, are a good place to begin working on an answer. Why did Zirkle organize a polemic against Lysenko? Zirkle was an ardent Cold Warrior and anti-communist. Despite the fact that his records are archived in the American Philosophical Society in Philadelphia, it is still fair to say that Zirkle is at least as best remembered for these activities as he is for his work as a botanist and historian of science. Of course Zirkle wanted to write history. That is the best way to influence the present.

December 4–5, 2009 I organized the International Workshop on Lysenkoism at the Graduate Center of the City University of New York and the Harriman Institute at Columbia University. The event brought together some 30 historians from nearly a dozen countries to present case studies on the impact of impact of, and reaction to,

[5]See Conway Zirkle, *Evolution, Marxian Biology and the Social Science* (Philadelphia, PA: University of Pennsylvania Press, 1959); Conway Zirkle, "The Early History of the Idea of the Inheritance of Acquired Characters and of Pangenesis," *Transactions of the American Philosophical Society* 35, no. 2 (1946): 91–151.; Garland Allen, *Thomas Hunt Morgan: The Man and His Science* (Princeton, NJ: Princeton University Press, 1978).

[6] See William deJong-Lambert, "The Uses of the Dead in the Science of Life: A Thanatology of Lysenkoism," *Studies in the History of Biology* 3, no. 2 (2011): 97–108. One detail I did not men tion above is that when I first investigated Zirkle's revisions to the Lamarckism entry in the 1961 *Encyclopedia Britannica*, I discovered (thanks to Richard W. Burkhard, Emeritus Professor, Department of History, University of Illinois. E-mail correspondence June 11, 2010) that someone had literally sliced them out of the pages in the editions found in the U.S. Library of Congress.

[7] The term "Lysenko affair" refers almost exclusively to events following the 1948 VASKhNIL conference up until Lysenko's formal denouncement in 1965. The terms "Lysenkoism" and "Lysenko affair" are currently being re-examined by historians of science. I will not continue to place them in quotation marks after their first use in this manuscript, but they should be understood as relics of a framework in which the Lysenko controversy was interpreted which is now obsolete. As Nikolai Krementsov has pointed out, the former term first appeared in 1945 in article in *Scientific Monthly*. See William deJong-Lambert and Nikolai Krementsov, "On Labels and Issues: The Lysenko Controversy and the Cold War," *The Jourannal of the History of Biology* 45, no. 1 (2012).

the Lysenko controversy in Italy, China, Japan, Mexico, Holland, East Germany, West Germany, Poland, Hungary, Czechoslovakia, the United States and the Soviet Union. In 2005 I had defended my dissertation, *The New Biology: Lysenkoism in Poland*, inspired by Nikolai Krementov's chapter on the import and export of Lysenkoism, published in Michael David-Fox and György Péteri's edited collection, *Academia in Upheaval: Origins, Transfers, and Transformations of the Communist Academic Regime in Russia and East Central Europe*.[8] As I conducted my research I became aware that there was, as yet, little written on Lysenkoism outside of the Soviet Union, and that broadening the scope of national case studies would be essential to moving the topic forward.

The idea of holding the workshop originated at a eugenics workshop organized by Magdalena Gawin at Warsaw University, "Eugenics, Modernisation and Biopolitics," April 17–19, 2008. I presented a paper on Polish geneticist Stanisław Skowron's twin experiences with eugenics and Lysenkoism, and after my talk another presenter, Francesco Cassata, informed me of his forthcoming work on Lysenkoism in Italy.[9] I also spoke with Michal Simunek about research he and his colleagues at Charles University were conducting on the Czech reaction to Lysenko. At this point it became clear that a network of Lysenko scholars was forming, and that a comparative project on Lysenkoism—of which the 2009 workshop was the first part—would be possible.

This book is a small contribution to that effort, and will hopefully prove worthwhile to historians of science who wish to guide their students through the basics of Lysenkoism, and outline the reaction in the United States. Though the workshop was a significant step forward in broadening the scope of research on the Lysenko controversy, most of the historiography still consists of analyses of Lysenko's career in the Soviet Union.[10] The response in the United States is a good place to begin

---

[8] William deJong-Lambert, "The New Biology: Lysenkoism in Poland" (Ph.D. dissertation, Columbia University, 2005). The dissertation has also been published by VDM Verlag; Nikolai Krementsov, "Lysenkoism in Europe: Export–Import of the Soviet Model," in *Academia in Upheaval: Origins, Transfers, and Transformations of the Communist Academic Regime in Russia and East Central Europe,* eds. Michael David-Fox and György Péteri's (Westport, CT: Bergin and Garvey, 2000).

[9] Francesco Cassatta, *Le due scienze. Il "caso Lysenko" in Italia* (Turin, Italy: Bollati Boringhieri, 2008).

[10] The standard bibliography when it comes to comprehensive accounts includes Conway Zirkle, *Death of a Science in Russia* (Philadelphia, PA: University of Pennsylvania Press, 1949); Julian Huxley, *Heredity East and West: Lysenko and World Science* (New York: Henry Shuman, 1949); David Joravsky, *The Lysenko Affair* (Chicago: The University of Chicago Press, 1970); Zhores Medvedev, *The Rise and Fall of T.D. Lysenko*, trans. Isadore Michael Lerner (New York: Doubleday and Co., 1971); Mark Adams, "Genetics and the Soviet Scientific Community, 1948–1965" (Ph.D. dissertation, Harvard University, 1972); Chapter 6 of Loren Graham's, *Science and Philosophy in the Soviet Union* (New York: Knopf, 1972); Dominique Lecourt, *Proletarian Science? The Case of Lysenko*, trans. Ben Brewster (London: NLB, 1977); V.A. Soyfer, *T.D. Lysenko and the Tragedy of Soviet Science*, trans. Leo Gruliow and Rebecca Gruliow (New Brunswick, NJ: Rutgers University Press, 1994); Krementsov, *Stalinist Science*; and Nils Roll-Hansen, *The Lysenko Effect: The Politics of Science* (New York: Prometheus Books, 2004). The works by Zirkle and Huxley are now treated as primary sources by historians of the Lysenko controversy.

when analyzing the international dimensions of the Lysenko affair, not only because the U.S. and the Soviet Union were the principle rivals in the Cold War, but also because Lysenko's anti-genetics campaign provoked the most controversy among scientists in its primary enemy.[11] Part of this was due to concurrent trends in U.S. Cold War culture, such as McCarthyism and loyalty investigations of the atomic scientists and other academics.[12] Biologists were in a delicate position due to the fact that their public stance on Lysenkoism was viewed through the lens of anti-communist propaganda, as well as the belief of many of their colleagues that scientists should not engage in politics.[13]

A large number of biologists in the U.S. engaged in the Lysenko controversy, however this book deals almost exclusively with four—Hermann J. Muller, Leslie Clarence Dunn, Theodosius Dobzhansky and Conway Zirkle.[14] The book also

---

[11] This was not necessarily the case elsewhere. In West Germany, for example, the Lysenko controversy was overshadowed by the importance of local upheavals in science in academia—particularly the founding of the Free University of Berlin. Alexander Schwerin, "Lysenkoism and the Reform of Postwar German Genetics" (paper presentation, The International Workshop on Lysenkoism, December 5, 2009).

[12] See, for example, Jessica Wang, *American Scientists in an Age of Anxiety: Scientists, Anticommunism, and the Cold War* (Chapel Hill, NC: University of North Carolina Press, 1999); Zuoyue Wang, *In Sputnik's Shadow: The President's Science Advisory Committee and Cold War America* (New Brunswick, NJ: Rutgers University Press, 2008); and Alfred K. Mann, *For Better or For Worse: The Marriage of Science and Government in the United States* (New York: Columbia University Press, 2000).

[13] See Audra Jayne Wolfe, *Speaking for Nature and Nation: Biologists as Public Intellectuals in Cold War Culture* (Ph.D. dissertation, University of Pennsylvania, 2002); Audra Jayne Wolfe, "What It Means to Go Public: The American Response to Lysenkoism, Reconsidered," *Historical Studies in the Natural Sciences* 40 (2010): 48–78; and Rena Seyla, "Defending Scientific Freedom and Democracy: The Genetics Society of America's Response to Lysenko," *Journal of the History of Biology* 45, no. 1 (2012).

[14] Dobzhansky, of course, was Russian, however by the early 1930s he had decided to remain in the United States; Other prominent figures in the response to the Lysenko affair in the United States include Milislav Demerec, Robert C. Cook, Karl Sax, Tracy Sonneborn, Salvador Luria and Isadore Michael Lerner. [See Krementsov, *International Science,* as well as R.C. Cook, "The Genetics Congress," *Journal of Heredity* 28 (1937): 24–6; Robert C. Cook, "Walpurgis Week in the Soviet Union," *The Scientific Monthly* 68, no. 6 (1949); *Journal of Heredity* 40, no. 7 (1949). In the latter case, the entire issue of the journal was devoted to the Lysenko controversy, as described in Chapter 11; For Sonneborn and Sax see, T.M. Sonneborn, "Heredity, Environment and Politics," *Science* 111, no. 2890 (1950): 529–39, as well as Karl Sax's response to Leslie Clarence Dunn's positive 1944 appraisal of Soviet Science (L.C. Dunn, "Science in the USSR: Soviet Biology," *Science* 99, no. 2561 (1944): 65–7.), in Karl Sax, "Soviet Biology," *Science* 99, no. 2572, (1944):298–9. Sax's negative assessment was then later contradicted by Anton Zhebrak (Anton Zhebrak, "Soviet Biology," *Science* 102, no. 2649 (1945):357–8). These were all later reprinted in Conway Zirkle's, *Death of a Science in Russia* (Philadelphia, PA: University of Pennsylvania Press, 1949); Salvador Luria wrote the letter to J.B.S. Haldane asking his cooperation in criticizing Lysenko, which I describe below. Lerner was later responsible for the English translation of Medvedev, *The Rise and Fall of T.D. Lysenko*]. In England other prominent supporters and critics were John D. Bernal, James Fyfe, C.D. Darlington, Sir Henry Dale and John Langdon-Davies [see John Langdon Davies, *Russia Puts the Clock Back: A Study of Soviet Science and Some British*

discusses the role of two British geneticists, Julian Huxley and J.B.S. Haldane, but only in terms of how their actions influenced their U.S. colleagues.[15] I devote most of my attention to these six individuals for two reasons:

1. They were among those most actively engaged in the controversy over the longest period of time.
2. Their reactions display the polarity of responses the Lysenko affair provoked.

Muller and Haldane's reactions to Lysenko have so far received the most attention, whereas the responses of Dunn, Dobzhansky, Huxley, and Zirkle are less well-known.[16] As will be described below, Muller was initially very enthusiastic about the future of Soviet genetics, not to mention the Soviet socialist system. His experiences in the Soviet Union in the 1930s, however, convinced him he had been wrong, and by the time the Lysenko affair got underway Muller had gone the opposite direction, and become a vocal critic of the USSR. Haldane is a prominent figure in the controversy due to the fact that he was the only geneticist to come out in support of Lysenko, thus highlighting the problem scientists faced when their scientific knowledge clashed with their pride or their political beliefs. Dunn and Dobzhansky initially collaborated against Lysenko, however soon recognized that their views on many subjects—politics in particular—were incompatible. Dobzhansky ultimately became frustrated by Dunn's reluctance to criticize Lysenko as harshly as he would have liked, a situation exacerbated by the fact that Dunn sacrificed his scientific career by devoting too much of his time to liberal political causes.

Huxley's reaction was due to an interest in Russian genetics that dated back to the early 1930s, but was even more so the product of his friendship with Muller. Huxley came to Muller's defense at a critical period in the controversy, when Muller's harsh criticism of Lysenko led many to accuse him of failing to show the same "scientific objectivity" he found lacking in Lysenko. Zirkle (the only one who

---

*Scientists* (London: Victor Gallancz Ltd., 1949); J.D. Bernal, "The Biological Controversy in the Soviet Union and Its Implications," *The Modern Quarterly* 4, no. 3 (1949):203–17, as well as J.L. Fyfe, "The Soviet Genetics Controversy," *Modern Quarterly* 3 (1948): 348; and James Fyfe, *Lysenko is Right* (London: Lawrence & Wishart, 1950); For Darlington see Oren Solomon Harman, "C.D. Darlington and the British and American Reaction to Lysenko and the Soviet Conception of Science," *Journal of the History of Biology* 36, no. 2 (2003): 309–52, and Oren Solomon Harman, *The Man Who Invented the Chromosome: A Life of Cyril Darlington* (Cambridge, MA: Harvard University Press, 2004)].

[15] The impact of the Lysenko controversy in Great Britain is a subject which deserves a book of its own and is beyond the scope of this present work. I also touch on some details of the French reaction, particularly the response of Louis Aragon, but only in terms of its influence upon events in the United States.

[16] For an account of Dobzhansky, Dunn and Muller's reactions to Lysenko see William deJong-Lambert, "Hermann J. Muller, Theodosius Dobzhansky, Leslie Clarence Dunn and the Reaction to Lysenkoism in the United States," *Journal of Cold War Studies*, (*forthcoming, summer 2012*). For an account of Dunn and Dobzhansky's translation of Lysenko's work *Heredity and Its Variability* within the larger issue of "pseudoscience," see Michal Gordon, "How Lysenkoism Became Pseudoscience: Dobzhansky to Velikovsky," *Journal of the History of Biology* 45, no. 1 (2012).

was not a geneticist) was deeply anti-Soviet. Of the six, Zirkle was most motivated by political considerations, as well as something else that will probably seem a little bit surprising. In this sense his behavior mirrors more precisely than most the activities of Lysenko and his supporters in the USSR.

The text which follows is intended to introduce students to the Lysenko controversy by presenting the development of Lysenko's career, and his impact upon Cold War science and culture in the U.S. They are arranged both thematically and chronologically, and designed to encourage students to ask questions, and debate the significance of, the various events that are outlined. Though this book provides useful data for historians wishing to make their own interpretations, it has been primarily written as a resource for use with advanced undergraduate and graduate students to get them to express their own opinions. One example of an issue they could address is how political events affect the development of scientific theories. This book does not provide an answer, or even present a hypothesis. It will hopefully encourage students to do this for themselves.

Chapter 1 introduces most of the individuals discussed in the text—Trofim D. Lysenko, Ivan Michurin, Nikolai Vavilov, Theodosius Dobzhansky, Julian Huxley, Leslie Clarence Dunn and J.B.S. Haldane. The visits to the Soviet Union of latter three are also described, as well as Lysenko's rise to prominence in the Soviet Union.

Chapter 2 describes Muller's emigration to the Soviet Union, the attack on genetics launched by Lysenko and his allies, and the reasons for Muller's exit. The VI International Genetics Congress in Ithaca, New York, and the circumstances surrounding the cancellation of the congress which was to have been held in Moscow, are also recounted. The chapter ends with Huxley's 1945 trip to Moscow, and the ambiguous situation in Soviet biology after World War II.

Chapter 3 discusses P.S. Hudson and R.H. Richens *The New Genetics in the Soviet Union*, Dunn and Dobzhansky's initial efforts to undermine Lysenko by translating his work into English, and the reaction of their colleagues. The chapter also sets up the rift between Dunn and Dobzhansky's views of the controversy, which will widen as the story progresses. The chapter concludes with a description of the 1948 VASKhNIL session where Lysenko launched his anti-genetics campaign.

Chapter 4 describes the initial reaction of the VASKhNIL session among scientists and the press. The chapter also recounts Muller's attempts to counteract Lysenko by publishing articles in a popular publication, *The Saturday Review of Literature*, and how Muller's views were received by readers. Conway Zirkle, and his anti-Lysenko work, *Death of a Science in Russia*, are also examined, as well as the developing "Red Scare" in the U.S.

Chapter 5 describes the wider reaction to the controversy, as a feature of Cold War culture in the U.S. Among the most important of these is the "Spitzer Affair," wherein a chemist at Oregon State University, Ralph Spitzer, was dismissed for teaching Lysenko's theories. I also write about Julian Huxley's book on the controversy, *Heredity East and West*, as well as a special issue of *The Journal of Heredity* devoted to the controversy.

Chapter 6 describes Lysenko's downfall, and the progress/decline of his critics. The chapter also focuses on Conway Zirkle's 1959 publication, *Evolution, Marxian Biology and the Social Scene*, a key text—for reasons that may surprise you—in the controversy. This then leads to the importance of biological theories of race, and the attempted revival of eugenics, a quarter century after World War II. The chapter is followed by an epilogue.

The sections marked off with asteryxes at the end of each chapter contain questions for classroom discussion. These questions are included to facilitate use of the text as a tool for teaching students about the Lysenko affair, and addressing the important issues the history of the controversy raises. They could also be used as the basis for research papers, or other types of projects. Depending upon the focus of the course (i.e., history or philosophy of biology, sociology of science, the Cold War), some questions will be more useful than others. The book assumes a knowledge of contemporary topics in the history of biology—particularly the wave of anti-Darwinism in the U.S., a drama best exemplified by the Scopes Trial—as well as the eugenics movement. It also assumes a broad familiarity with U.S. Soviet relations in the early to mid-period of the Cold War.

# Chapter 1
# Mendelist-Morganists and Michurinism

*...changes in the nature of the ground produce changes in animals; for instance, the water snake changes into a viper, if the marshes are dried up.*

—Theophrastus, c. 371–287 B.C.E.[1]

## 1.1 Trofim D. Lysenko and Theodosius Dobzhansky

In 1927 a grain crisis and the drive to industrialize in the Soviet Union led to war on the peasantry. As collectivization continued, kulaks were deported, exiled, arrested, and shot in a frenzied campaign that was only moderated once Stalin declared "dizziness from success." The assault was soon renewed as the Communist Party set production quotas far above what was possible, and requisitioned grain from starving peasants. By the early 1930s famine in the Ukrainian breadbasket was so severe that the sight of victims dropping dead on the street from hunger ceased to provoke notice. Cannibalism was so widespread that signs had to be posted admonishing: "Eating ones own children is an act of barbarity."[2] Meanwhile, the Soviet government asked biologists what their science had to offer for increasing agricultural productivity. This was the environment in which Trofim Denisovich Lysenko began his career.

---

[1] Quoted in L.P. Coonen, "Lysenkoism in Athens," *Science* 119, no. 3098 (1954): 694–5.

[2] For a background of Stalin's declaration in the broader context of the collectivization drive, see Chapter 6 of Robert W. Davies, *The Socialist Offensive: The Collectivization of Soviet Agriculture* (Cambridge, MA: Harvard University Press, 1980), pp. 269–310; See also Robert Conquest, *The Harvest of Sorrow: Soviet Collectivisation and the Terror-Famine* (London: Pimlico, 1986); and Lynne Viola and others, eds., *The War Against the Peasantry, 1927–1930: The Tragedy of the Soviet Countryside*, trans. Steven Shabad (New Haven, CT and London: Yale University Press, 2005). A photograph of the sign referred to can be found in Stéphane Courtois et al., *The Black Book of Communism: Crimes, Terror, Repression*, trans. Jonathan Murphy and Mark Kramer (Cambridge, MA: Harvard University Press, 1999).

W. deJong-Lambert, *The Cold War Politics of Genetic Research: An Introduction to the Lysenko Affair*, Archimedes 24, DOI 10.1007/978-94-007-2840-0_1,

In 1927, T.D. Lysenko's name first appeared in the Soviet press. A feature story in *Pravda* described:

> If one is to judge a man by first impression, Lysenko gives one the feeling of a toothache; God give him health, he has a dejected mien. Stingy of words and insignificant of face is he; all one remembers is his sullen look creeping along the earth as if, at very least, he were ready to do someone in.[3]

The article portrayed Lysenko as a peasant practitioner whose skills had gained the respect of scientific experts. He was a "barefoot scientist" who solved complex problems on a "little old slip of paper," and only smiled at the mention of Poltava cherry dumplings with sugar and sour cream. According to *Pravda*, his name was becoming known, he had followers and pupils, and in the winter Lysenko was visited by agronomic luminaries who stood before the green fields of his experiment station, gratefully shaking his hand.[4]

This early, heroic image of Lysenko contrasts starkly with descriptions which would be given just over 20 years later in the West by critics alarmed by the authority granted to him in Soviet biology. Terms such as "charlatan" and "quack," along with frequent comparisons to Savonarola and Rasputin, would affirm Lysenko's reputation as an ignorant fanatic whose success was proof that Soviet science had been smothered by totalitarianism and corrupted by Marxist ideology.[5] Supporters, on the other hand, would cite Lysenko as evidence of the revolutionary power of Soviet socialism and the triumph of the belief that science should serve the common good and provide tangible benefits.

Lysenko's significance, however, was actually far more complicated. By effectively banning genetics in the Eastern Bloc he engaged in one half of the "struggle for Darwin's legacy." Lysenko also forced biologists in the West to confront the political implications of their role, at the same time as they attempted to challenge the views of someone few regarded as a scientist. Because "Lysenkoism" has become a synonym for pseudoscience, it is important to ask how this happened and why.[6]

Lysenko was born in 1898 in the Ukraine. His parents were peasants and as the eldest son he was expected to help work in the fields. Lysenko did not learn to read or write until sent to the village school at age 13. He applied to study at an agricultural

---

[3] Medvedev, *The Rise and Fall of T.D. Lysenko*, pp. 11–2. For an alternative translation of this passage see Joravsky, *The Lysenko Affair*, p. 59.

[4] Ibid.

[5] Girolamo Savonarola (1452–1498) was a Dominican priest and leader in Florence famous for his anti-Renaissance stance and extremist efforts in religious reform; Grigori Rasputin (1869–1916) was a Russian holy man, decried by his critics as a religious charlatan whose influence played a role in discrediting the Romanovs prior to the Bolshevik Revolution.

[6] For a discussion on "Lysenkoism" as "pseudoscience," see Jenny Leigh Smith, "Lysenko's Legacy: Ignorance, Bliss and the Persistence of People's Science." Paper presentation at the International Workshop on Lysenkoism, December 4, 2009. Available online at http://www.youtube.com/watch?v=37ck9UPiJc8. See also Gordon, "How Lysenkoism Became Pseudoscience: Dobzhansky to Velikovsky."

college in 1916 but failed the requisite examination in Scripture. Lysenko tried again and succeeded, but classes were continually interrupted by the turmoil of Revolution and Civil War. The town where he studied changed hands several times between the Red and White armies. Lysenko benefited from the desire of the Bolshevik leaders to educate those who had in the past been left uneducated—the children of the peasant and proletariat. However, Lysenko never completely compensated for the turmoil of those early years. He was quick and clever, but did not—according another chronicler of the Lysenko controversy, Valery Soyfer— possess a "love of knowledge for its own sake"—"the drives that ordinarily inspire scientists." Lysenko remained willfully ignorant of research outside of the Soviet Union, hated math and never passed a formal scientific examination.[7]

Lysenko also never learned a foreign language and spoke Russian with a thick accent and without regard to grammar. Detractors enlisted these details against him, but to others this was proof of the dirt under his fingernails, credibility as he emerged out of a movement of peasant scientists working in the fields. Although progressive agricultural reform was the goal, the hut labs were more often the site of strange and unlikely experiments—soaking seeds in salt or the juice of dung to get them to grow better. Here it seems Lysenko's ignorance and intolerance of dissent were assets.[8]

In 1927 biologist Theodosius Dobzhansky left the Soviet Union on a Rockefeller grant, never to return. Dobzhansky began his career in T.H. Morgan's lab at Columbia University in the United States, and would go on to become one of the most famous figures in genetic research in the twentieth century. He was, however, reviled in his own country as a traitorous non-returner and a "*mrakobes*"—an obscurantist or "devil" who spreads darkness. In an oral history recorded in 1962, Dobzhansky recounted his development as a biologist in the Soviet Union in the 1920s and the reasons he never returned. He also detailed the impact Lysenko had on the lives and careers of those around him. Though Dobzhansky was at the center of the effort to criticize and undermine Lysenko in the United States after the Second World War, he never let the effort distract him from his work as a scientist.

Dobzhansky was born about a year after Lysenko. His father came from Polish nobility who lost their land after an uprising against Russian rule in 1863.[9] Dobzhansky graduated high school in 1917, just after the February Revolution. He was just 12 days too young to be drafted,[10] and attended the University of Kiev.

[7] Valery Soyfer, *T.D. Lysenko and the Tragedy of Soviet Science*, trans. Leo Gruliow and Rebecca Gruliow (New Brunswick, NJ: Rutgers University Press, 1994), p. 9.

[8] The physical descriptions of Lysenko and details of his background come from Wacław Gajewski, "Lysenkoism in Poland," *Quarterly Review of Biology* 65, no. 4 (1990): 423–34; Szczepan Pieniążek, *Pamiętnik sadownika* (Warszawa, Poland: Rozwój SGGW, 1997); Medvedev, *The Rise and Fall of T.D. Lysenko*; Soyfer, *T.D. Lysenko and the Tragedy of Soviet Science*, pp. 7–9; Joravsky, *The Lysenko Affair*.

[9] The Reminiscences of Theodosius Dobzhansky, Oral History Research Office. Columbia University, 1962, pp. 1–2.

[10] Sophia Dobzhansky Coe, "Theodosius Dobzhansky: A Family Story," in *The Evolution of Theodosius Dobzhansky*, ed. Mark B. Adams, (Princeton, NJ: Princeton University Press, 1994), p. 13; The Reminiscences of Theodosius Dobzhansky, p. 40.

Kiev was close to the front lines so Dobzhansky witnessed things going from bad to worse—shortages, soldiers deserting, the Germans advancing. By the October Revolution Dobzhansky knew almost no one who favored the Lenin and the Bolsheviks.[11] University life still carried on though, and in the spring of 1917 the Dnieper River flooded. Insects and other small animals caught in the deluge rose to the surface as floating debris, where the wind drove them to the bank. Among the vast number of insects delivered by the flood, Dobzhansky discovered a new member of the ladybird beetle family. This enabled him to publish his first scientific paper in 1918.[12] As had been the case with Darwin, Dobzhansky's fascination with beetles would draw him towards larger questions concerning what constitutes a species, and how they evolve.[13]

The situation in Kiev became increasingly chaotic as Europe reacted to the revolution in Russia and the outcome of the First World War. In March, 1919, Ukrainian Bolsheviks, with Russian support, founded the Ukrainian Soviet Socialist Republic. Once the turmoil of war, revolution, and counter-revolution was over, Dobzhansky became an instructor at the agricultural institute near Kiev where Lysenko was a student.[14] Dobzhansky's comments in his oral history on the intellectual milieu in which Lysenko was educated are useful for the understanding the later trajectory of his career. It was also during this period that Dobzhansky read an account of the work of the work being done by T.H. Morgan and his students in the "fly room" at Columbia University in the United States. The senior biologists he knew believed Morgan's *Drosophila* mutants were monstrosities, with no practical relevance to the study of evolution. But Dobzhansky and others understood the importance of Morgan's genetics. The more he discussed Morgan's work with his contemporaries the clearer it became to Dobzhansky that he must pursue it.

In Kiev Dobzhansky received little encouragement for his work, so he made periodic pilgrimages to Moscow and Petrograd to read the works of leading Russian geneticist, Nikolai Vavilov, as well as the books Vavilov had brought back from the West. In Petrograd, Dobzhansky and other aspiring biologists stayed in a "house of scientists," established by the Bolsheviks in the former palace of a grand duke. For a token fee, they could work and discuss what they discovered.[15]

Dobzhansky wanted to work with *Drosophila melanogaster*, but if he stayed in Kiev he would be on his own. Finally, after publishing the results of research he had done with *Drosophila* samples in Moscow, he received an offer from one of the

---

[11] The Reminiscences of Theodosius Dobzhansky, pp. 47–8.

[12] The Reminiscences of Theodosius Dobzhansky, pp. 52–4.

[13] See R.C. Fisher, "An Inordinate Fondness for Beetles," *Biological Journal of the Linnean Society* 35 (1988): 313–9; Nikolai Krementsov, "Dobzhansky and Russian Entomology: The Origin of His Ideas on Species and Speciation," and Mikhail Konashev, "From the Archives: Dobzhansky in Kiev and Leningrad," in *The Evolution of Theodosius Dobzhanksy,* ed. Mark Adams (Princeton, NJ: Princeton University Press, 1994).

[14] The Reminiscences of Theodosius Dobzhansky, p. 5.

[15] The Reminiscences of Theodosius Dobzhansky, p. 122.

leading figures in Soviet genetics, Iurii Filipchenko, to work in Petrograd. Dobzhansky's wife soon joined him. In the summer they traveled to Central Asia for research, and when back in Leningrad, Dobzhansky continued his work with *Drosophila melanogaster*. When he went to the United States on a Rockefeller grant to work in Morgan's lab it was expected Dobzhansky would return to lead *Drosophila* research in the Soviet Union.[16] But he never did.

## 1.2 Lenin, Burbank, Michurin and Vavilov

The perennial problem of food production in Russia was worsened by the turmoil of war and revolution. The result was crippling famine. A popular book describing the wonders achieved by Luther Burbank in the American West captured Lenin's imagination.[17] In the United States Burbank was celebrated as the Thomas Edison of agriculture, the Henry Ford of flowers. He was also, as one of the most controversial figures in the history of plant breeding, alternately praised and criticized by a wide range of critics and supporters.[18] From his ranch in Santa Rosa, California, Burbank brought forth new varieties of fruits, nuts, grains, grasses, vegetables and flowers, including Royal walnuts, spineless cactus, Japanese golden mayberry, Perfection plums, potato-tomato hybrids, the Nicotunia—bred by crossing nicotianas with petunias—and a flowering quince he called "Dazzle."[19] Burbank grafted and cross-pollinated obsessively, selling his creations for hundreds and thousands of dollars through the wonders of advertising: "The *best of millions* of cross-bred, hybrid and seedling plants, which are now and have been produced at the rate of a million or more a year," according to his catalogue.[20] Unlike Gregor Mendel, Burbank did not care about statistics or predictable ratios. Rather, like I.V. Michurin—who Lenin would soon discover—Burbank believed that environment (in his case, the frontier climate of Southern California) could enable the talented practitioner to transform nature.

Burbank also believed that what he knew about breeding plants could easily be applied to mankind, and some of his notions reflected the contemporary enthusiasm for eugenics. Burbank published a popular treatise in *Century* magazine, "The Training of the Human Plant," proposing that the variety of immigrants coming to the United States offered a grand opportunity to apply the ineffable magic of crossing

---

[16] Konashev, "From the Archives: Dobzhanksy in Kiev and Leningrad," p. 78.

[17] B: Z67 Conway Zirkle Papers, Dobzhansky, T. End of Genetics in the Soviet Union, p. 5. The American Philosophical Society.

[18] See Walter L. Howard, *Luther Burbank: A Victim of Hero Worship* (Waltham, MA: The Chronica Botanica Co., 1945).

[19] Peter Dreyer, *A Gardener Touched with Genius: The Life of Luther Burbank* (Berkeley and Los Angeles: University of California Press, 1985), pp. 91–105.

[20] Dreyer, *A Gardener Touched with Genius*, p. 98.

species. However careless selection, he cautioned, could result in vast harm. It is bad enough, he argued, to take a plant that is poisonous and cross it with another that is not, thus making the wholesome plant evil. But if we blend two poisonous plants to make a third more virulent—a vegetable degenerate—and send its decadent descendants adrift to multiply, are we not "foes" of the human "race?" Is it not even worse, Burbank asserted, to allow impaired people to have children?[21]

Numerous geneticists visited Burbank, curious about his work. In 1906 the Carnegie Institution sent George H. Shull to Burbank's nurseries to see if his work had practical value, and if he had developed methods that could be applied to mass production.[22] Shull's verdict was "no." Burbank's curiosities were the product of intuition and guess-work, and Burbank himself more a P.T. Barnum than a Benjamin Franklin.[23] But that did not matter—to the wider public Burbank was a great scientist. President Franklin D. Roosevelt honored Burbank by making him the first and only "geneticist" ever to appear on a postage stamp until Barbara McClintock was similarly honored in 2005.[24] But Burbank's work had nothing to do with genetics.[25]

Nikolai Vavilov was among those who made the pilgrimage to Burbank's farm. Lenin had charged Vavilov with the responsibility for organizing an institute for genetics and plant breeding to end the chronic problem of insufficient food production in Russia. Vavilov shared the Bolshevik belief that communism made possible the development of science on a scale capitalist countries could only dream about. To this end he traveled the globe collecting plant samples and a library of literature on the biological sciences. Vavilov knew 22 languages and studied archaeological digs to learn the source of cultivated plants. He hoped to create a periodic table of the plant word, providing botanists and plant breeders with the same possibilities as chemists and engineers.[26]

Vavilov slept four hours a night and commuted two nights a week between Moscow and Petrograd. He occupied an English-style country house, formerly belonging to the Imperial family, and on the grounds Vavilov constructed an extensive series of greenhouses, laboratory buildings and experimental fields. From his

---

[21] Ibid, pp. 252; These views were expressed in "The Training of the Human Plant," an essay he published in *Century* magazine in 1907.

[22] Bentley Glass and George H. Shull, "The Strange Encounter of Luther Burbank and George Harrison Shull," *Proceedings of the American Philosophical Society* 124, no. 2 (1980).

[23] Ibid, pp. 14, 227–8.

[24] Ibid, p. 223.

[25] In his biography of Burbank, Peter Dreyer describes the relationship between Burbank's career and the Lysenko affair: "Decades later Burbank's name was appropriated by T.D. Lysenko and his followers, then engaged in ruthlessly dismantling the entire structure of Russian genetic science. ... None of this had anything to do with Burbank. But he had been branded 'unscientific' and adopted into the bargain in Russia as a Lysenkoist totem to set beside the canonized Michurin. Accordingly, in Western eyes, he became a kind of Lysenkoist by posthumous association." Dreyer, *A Gardener Touched with Genius*, pp. 221–3.

[26] Soyfer, p. 46.

Institute of Applied Botany he organized a chain of agricultural experiment stations, stretching from Murmansk on the Polar Sea, down to the southern Caucasus, and from the Western border of the Ukraine to Eastern Siberia.[27]

Vavilov's charisma made his enthusiasm contagious. He had a flattering memory for names and personal details, and a talent for pulling from his stuffed briefcase the exact piece of information required to answer any pressing question at hand.[28] Dobzhansky later described Vavilov's institute as a "beehive" of activity.[29] Specimens of domesticated plants were arranged in dining rooms decorated with tiles and paintings portraying abundant feasts—property of the previous owner. Maps of Vavilov's expeditions hung over the scarlet and gilt of the old drawing room, which now served as his study. Vavilov worked at a long desk covered in plants and papers, with a small area cleared for cake and tea. Glasses were served every few minutes, old ones replaced by new even if they had not been drunk.[30]

Part of Vavilov's work involved investigating and evaluating efforts to improve Soviet agriculture. To that end, in 1920, he visited the nursery of an aged and eccentric plant breeder, Ivan Vladimirovich Michurin. Lenin had been told there was a "Russian Burbank" doing important work that deserved attention.[31] The breeder's work might prove useful, though the peasants claimed it was the work of a "magician." Vavilov was impressed with the old man's notes and suggested he deserved some modest support. This was all the encouragement Michurin needed.[32]

Later images of I.V. Michurin portray a humble man with a short beard and tired eyes staring off elsewhere, as though barely noticing the medals on his chest. Renowned Soviet filmmaker Alexander Dovzhenko made a movie about him, which was released in 1949. The picture portrayed Michurin as a man who struggled until his work was finally appreciated in the revolutionary environment of Soviet socialism. He was presented as a stubborn cultivator of fruit trees, whose tireless efforts were ignored by the corrupt, tsarist bureaucracy, yet received the full support of the Bolsheviks after the Revolution. The film also depicted him fending off attempts by capitalists from the United States to bribe him into working for them.[33]

In fact, Michurin was born into an impoverished noble family who had been reduced to working their own orchard to eat. His mother died of tuberculosis and his father was taken away to a madhouse for—as David Joravsky put it in his landmark

---

[27] David Joravsky, "The Vavilov Brothers," *Slavic Review* 24, no. 3 (1965): 381–94; Reminiscences, Part I. B: D65 Dobzhansky, Theodosius. The American Philosophical Society, pp. 166–7.

[28] Joravksy, "The Vavilov Brothers," p. 166.

[29] The Reminiscences of Theodosius Dobzhansky, p. 166.

[30] J.G. Crowther, *Soviet Science* (New York: E.P. Dutton and Company, 1936), p. 262.

[31] Theodosius Dobzhansky, "The End of Genetics in the Soviet Union," *Bulletin of the Atomic Scientists* 5, no. 5 (1949): 144; For further information on the roots of Michurinism in Russia, see Douglas Weiner, "The roots of 'Michurinism': Transformist biology and acclimatization as currents in the Russian life sciences," *Annals of Science* 42, no. 3 (1985): 243–60.

[32] Joravsky, *The Lysenko Affair*, pp. 46–7.

[33] Aleksander Dovzhenko, *Michurin*. In the United States the film was distributed under the title *Life in Bloom*.

work, *The Lysenko Affair*—"coming out with a dance instead of the dirge" at her funeral. Michurin was the only one of his seven siblings to survive.[34]

Michurin was mechanically inclined, and worked as a repairer of railroad signals and watches in the small town of Kozlov. He quit both jobs at the age of 34 and bought 34 acres with borrowed money to set up an orchard near the city of Tambov. He was determined to succeed where his father had failed, and set to attempting the creation of new varieties of fruit by grafting seedlings onto types he wished them to resemble.[35]

The temptations of capitalists, portrayed in Dovzhenko's film, actually amounted to no more than two visits from an American plant explorer who tried to arrange regular purchases from his nursery. The offer was withdrawn because Michurin demanded too much money, and even after he reduced his demands the project fell through. For Michurin, however, the story expanded for the rest of his life, until he had agents from the U. S. Department of Agriculture visiting him on a regular basis, begging him to emigrate to the United States. They ostensibly offered him a ship to carry his plant collection, the directorship of a plant-breeding station, 100 assistants and an annual salary of $32,000, at a time when the highest salary offered by the Department of Agriculture was $4,000.[36] But what did money mean to a patriotic Soviet?

Michurin concentrated on the creation of vegetative hybrids—melon and squash, peach and pear, over 100 varieties of apples—of dubious authenticity or value. He also developed and promulgated the "mentor method," based on the belief that young cuttings would acquire characteristics from the older, superior varieties they were grafted onto. Michurin believed maturity was the critical factor in determining which direction this transfer of characteristics would occur because, as he pointed out, it would be ridiculous to expect that a child would react to its environment in the same way as an adult. Like Burbank, anthropomorphism formed the basis for Michurin's understanding of the natural world, and he believed all living things were endowed with an intelligent ability to adapt in the struggle for existence. This infinite plasticity implied to Michurin a challenge not to wait on the kindness of nature, but to create new varieties even nature itself could not imagine. He dismissed scientists and academics who sought to understand problems to be answered by Mendel's statistics, calling them "caste priests of jabberology." Ignorance was mutual. Michurin spent many years attempting to receive financing and a position of authority from the Russian government, and then the Soviets.[37]

At the end of the First All-Russian Agricultural Exhibition in the fall of 1923, *Izvestia* published a headline, "Kozlov or Washington?" The story faithfully

[34] Joravsky, *The Lysenko Affair*, pp. 40–1.

[35] Ibid.

[36] Ibid, pp. 44–5.

[37] Ibid, pp. 42–4.

regurgitated Michurin's delusions of the fame awaiting him in the West, and within a month he was given the support he had always felt he deserved.[38] Michurin's seventieth birthday in 1925 was celebrated as a national holiday, and he acquired the moniker "father of apples" in acknowledgement of his own favored objects of experimentation. The town of Kozlov was renamed Michurinsk, and a statue of its namesake was erected. For Soviet journalists, Michurin became a figure who evinced the power of the common man to transform unschooled intuition into scientific breakthrough.[39]

Experts were more cautious, conceding only that Michurin's data could be useful if verified by scientists. But when scientists conducted rigorous examinations of "mentors" in Michurin's nursery, their findings were angrily rebuked. Michurin fumed that their performance was "slipshod" and accused them of "undermining faith" in what he had done. Vavilov was also dismissive of Michurin's work, and published an article describing Michurin's melon and squash hybrids as imaginary. Michurin's career, however, was to prove the power of imagination and resentful determination, underpinned by the same fantasy that powered Burbank's career in the U.S. Michurin's advocates insisted that his relationship with plants was so intimate that he could predict at a glance how a plant would develop and bear fruit in the future. Michurin's instinct supposedly enabled him to subject plants to his will. Thus the Soviets brushed science aside with an appeal to the authenticity of a lonely genius—a peasant, a patriot. It was a bitter myth Lysenko would eagerly latch onto.[40]

## 1.3 H.J. Muller

The work of T.H. Morgan and his students in the "fly room" at Columbia University is a fundamental part of the Lysenko story. Morgan's name, with the suffix –ism attached, would constitute one-third of the enemy ("Mendelism-Weismannism-Morganism") Lysenko and his followers positioned themselves against. The fact that fruit flies were the focus of research in Morgan's lab became a handy piece of evidence for those who wished to portray genetics as useless and impractical. The fact that one member of the fly room was Hermann J. Muller, did not help much either.

Muller started at Columbia on a fellowship at age 16, and went straight on for his Ph.D. He was 5 ft 2 in. tall, and made up for his short stature with an intense intellect and combative nature. Muller would routinely pace back and forth when absorbed in conversation, and was known to literally jump up and down when excited by a

[38] Ibid, pp. 49–50.

[39] Ibid, p. 49.

[40] Ibid, pp. 50–4.

new idea.[41] He also had a "priority complex," and his whole life bitterly resented those who he felt stole his discoveries.[42] At Columbia he developed an interest in eugenics and socialism, to compliment his burgeoning fascination with genetics. This heady combination would influence Muller—with at times disastrous consequences—for the rest of his career.

Muller's attraction to socialism made him suspicious of the agendas of elitism and prejudice which eugenics could (and did) serve. But he also believed the "less fit" constituted a genetic problem science should be allowed to solve. In addition to his own Biology Club, Muller joined the Intercollegiate Socialist Society, and the Peithologian Society. The latter was a literary and philosophical discussion group which derived its name from the Peitho, the goddess of amorous persuasion. Muller regaled them with his ambitions to use psychology, sociology, physiology, and genetics to breed a nobler race of human beings.[43] These ideas would later be published in a book Muller would eagerly present to Joseph Stalin, titled *Out of the Night*.

When he completed his Ph.D. in 1915, Muller joined the Biology Department at the newly-established William Marsh Rice Institute, later re-named Rice University, in Houston, Texas. He remained at Rice until 1918, then had a brief 2-year appointment back at Columbia which, to his disappointment, was not renewed.[44] Muller returned to Texas and took a position at the University in Austin, where his east coast politics would soon find him at odds with the local climate.

In 1922 Muller brought the first samples of *Drosophila melanogaster* to the Soviet Union. Upon returning to the United States Muller published an article describing his visit in *The Scientific Monthly*, as well as a partial list of biological research institutes and the work they conducted in *Science*.[45] Muller's motivation for visiting the Soviet Union and writing an article chronicling his experience was to reestablish communication between U.S. and Soviet biologists, and satisfy his curiosity about the outcome of the Bolshevik Revolution. The purpose of the list of institutes and research was to enable and encourage American scientists to send reprints and back numbers of periodicals, to help their Russian colleagues gain knowledge of the work being done in the West.[46]

---

[41] James Schwartz, *In Pursuit of the Gene: From Darwin to DNA* (Cambridge, MA: Harvard University Press, 2008), p. 194; Schwartz's book is one of two biographies written on Hermann J. Muller. The first, cited below, was written by Elof Axel Carlson. Carlson was a graduate student of Muller's, and wrote an authorized biography.

[42] Elof Axel Carlson, *Mendel's Legacy: The Origin of Classical Genetics* (Cold Spring Harbor, NY: Cold Spring Harbor Press, 2004), p. 206.

[43] Elof Axel Carlson, *Genes, Radiation and Society: The Life and Work of H.J. Muller* (Ithaca, NY: Cornell University Press, 1981), p. 35.

[44] Carlson, Genes, *Radiation and Society*, pp. 116–7.

[45] H.J. Muller, "Observations of Biological Science in Russia," *The Scientific Monthly* 16, no. 5 (1923): 539–52; H.J. Muller, "Partial List of Biological Institutes and Biologists Doing Experimental Work in Russia at the Present Time," *Science* LVII, no. 1477 (1923): 472–3.

[46] Muller, "Partial List of Biological Institutes," p. 472.

In the *Scientific Monthly* Muller portrayed Soviet biologists as an active scientific community whose greatest obstacle was keeping up with research being conducted beyond their own borders. This resulted in wasteful duplication—a situation which, in Muller's opinion, slowed progress in Western genetics as well.[47] Muller wrote it was important to let Soviet biologists know that they were welcome to publish in American journals, and also issue official invitations to enable them to visit. Muller cited a recent trip by Vavilov, at the invitation of the American Society of Phytopathologists, as an example of the impact such exchanges could have:

> Professor Vavilov ... on returning to his country, had made a special tour throughout the land, giving numerous illustrated addresses to interested audiences of thousands of persons on scientific work in America. And, on account of this, wherever I went, I found certain general viewpoints and salient features of our American research already familiar, even though the investigators might not have had a chance to read a single American paper.[48]

Muller offered vivid descriptions of the challenges Soviet scientists had faced since the Revolution—famine, over-crowding and insufficient equipment. Among the exceptions Muller depicted was Vavilov's institute. Muller included a photograph in the article showing a genetics station which was formerly the mansion of the Grand Duke Michael Nikolayevitch, presented to him by Queen Victoria. Muller also recorded that the "bizarre buildings of the monk Rasputin," granted to him by the tsar, were now being used for agricultural research.[49] Muller was impressed not only by the scale of the process of classifying the vast collection of varieties of plants and cereals, but also by the fact that they were being tested with a view towards practical use, as well as theoretical principles of variation.[50]

Muller also described an exchange he had with one of Vavilov's colleagues, L.S. Berg, to illustrate an important difference in official attitudes towards science in the United States and the Soviet Union. When Muller mentioned that some state legislatures in the U.S. were considering bills to outlaw the teaching of evolution, Berg countered that he was having problems getting his book cleared for publication by the Soviet government because they suspected that its content contradicted Darwin. [51] Muller clearly believed that the problem of a government disputing research because it did not conform to their interpretation of a scientific theory, was preferable to a situation in which the government attempted to ban the theory altogether, and replace it with religious doctrine.

Muller was deeply impressed by the relationship between science and society in the Soviet Union. His initial positive impression, in addition to the belief that socialist societies were better-prepared to implement scientific advances as official

[47] Ibid, p. 551.

[48] Ibid, p. 552.

[49] Ibid, p. 548.

[50] Ibid, p. 549.

[51] Ibid. A description of the controversy surrounding Berg's theory of nomogenesis can be found in David Joravsky, "Soviet Marxism and Biology Before Lysenko," *Journal of the History of Ideas* 20, no. 1 (1959): 85–104.

policy, would later inspire him to immigrate to the Soviet Union. His conflict with Lysenko, however, would cause him to realize that state support of science could easily translate into interference and control.

Muller married soon after his trip to the USSR. His wife, Jessie Marie Jacobs, was a math instructor at Austin. Their courtship began when she helped him with a formula he needed for his work on fly mutation. She had received her Ph.D. from the University of Chicago at a time when few women received doctorates, much less in math. Their son was born little more than a year later, in 1924. Muller was ecstatic, but Jessie was distraught—her career was over. The mathematics department terminated her appointment on the basis that she could not be expected to be both a good teacher and a mother at the same time.[52]

Muller's fears of an anti-evolution trend sweeping the nation were realized the following summer by the "Monkey Trial," when a high school teacher was fined $100 for teaching about Darwinian Evolution. Church based antipathy towards Natural Selection had declined by the early twentieth century, as Darwin's theory became less influential. However, the development of genetics—soon to be widely accepted as the scientific underpinning of evolution—provoked a renewed crusade against evolution.[53] For Muller and other biologists of his generation, such as Leslie Clarence Dunn, the Scopes Trial would serve as an example demonstrating that religious fundamentalism was as dangerous an enemy to scientific progress in the United States, as state control of science in the Soviet Union.

The year after the Scopes Trial Muller had the breakthrough that made him famous. He found that X-rays produce mutations in *Drosophila*. On November 3, 1926 Muller swept two male fruit flies from an examination plate into a gelatin tube that was perforated with air holes burned into it by a hot needle. A local radiologist helped him subject the flies to four doses of radiation. Muller then paired the flies with females, observed the mating dance and examined the offspring. The results were so remarkable that he literally jumped out of his chair. By now it was night and the only person to tell was a botanist working in the lab below. Muller opened his window and shouted down the good news. As he found more and more mutations Muller yelled out the window again and again: "I got another!"

Muller spent almost the entire next day and night in the lab, bringing in a cot to sleep. By the time the experiment was over less than two months later, his notebook was crammed with entries on over 100 mutations. This was half the number that had been found in the entire 16-year history of work on *Drosophila*.[54] It was an extraordinary result that earned him the Nobel Prize 20 years later.

Muller was also concerned, however, with the increased use of X-rays in medical practice, and began using his prominence as a podium to ask questions, regularly

---

[52] Ibid, pp. 133–4.

[53] Edward J. Larson, *Summer for the Gods: The Scopes Trial and America's Continuing Debate Over Science and Religion* (Cambridge, MA: Harvard University Press, 1998), pp. 19–23.

[54] Carlson, *Genes, Radiation and Society*, p. 145.

disappointing audiences that preferred to hear scientists say optimistic things. In 1928, just two years after his first successful experiments with radiation and *Drosophila*, Muller was invited to give an address to the Medical Society of Waco County at Baylor University. Muller warned his audience not to use radiation unnecessarily, or at the very least limit exposure. He countered talk of using X-rays to produce desirable mutations by pointing out the moral problem of selection. His audience grew angry and several people stalked out of the room. A follow-up address Muller was to give to a society of radiologists was cancelled.[55]

Nevertheless, Muller's work was extraordinary. It also caught the attention of a new member of Morgan's lab from the Soviet Union, Theodosius Dobzhansky. In Moscow Dobzhansky had worked on *Drosophila* using the mutant stocks Muller had delivered. Now he arrived a complete outsider, appalled by the filth of the fly room. Morgan's lab was a growing legend, but Dobzhansky disliked its provincial American character. The chatter was informal shoptalk and Dobzhansky came from a continent where science was a patrician duty, a higher cultural calling. He often remarked that in Europe evolution was not just a scientific topic, it was philosophical.

Morgan brought Dobzhansky out to California with him in 1928 to teach at Cal Tech, and continue work on *Drosophila*. While Dobzhansky was in California, Vavilov stopped in for a visit on his way back to the Soviet Union from Peru, where he had been researching potatoes in the Andes. A photograph taken at the time shows Dobzhansky in knee-high, lace-up boots, jodhpurs and a tie; to his left is his wife, and then Vavilov next to another geneticist, Georgii Karpechenko—both dressed in heavy suit jackets. They all smiled and squinted in the hot Southern California sun. Dobzhansky had a Model A Ford and they drove to Sequoia National Park and to view the agriculture of the Great Valley. They talked non-stop—or Vavilov did mostly—since Dobzhansky required more than four hours of sleep. Vavilov told Dobzhansky he wore a kind of "spectacles" that permitted him to see some things and not others.[56] He said Dobzhansky had nothing to fear and should return to the Soviet Union: "We have to ignore, we have to leave out of consideration, the political matters with which we do not agree." The advancement of science was all that mattered, he said, but Vavilov returned alone.[57]

Vavilov's statements concerning "political matters" which were best ignored referred to the politicized atmosphere which had developed in the Soviet Union around discussions concerning heredity and evolution since the middle of the 1920s. The Scopes Trial had given Soviet scientists cause to feel superior, particularly after one biologist writing on the trial characterized Tennessee as "one of the most

---

[55] Carlson, *Genes, Radiation and Society*, pp. 159–60. For more on Muller's skepticism on the use of X-rays, even in his own work, see H.J. Muller, "Radiation and Genetics," *American Naturalist* 64 (1930): 220–51.

[56] The Reminiscences of Theodosius Dobzhansky, p. 166.

[57] The Reminiscences of Theodosius Dobzhansky, pp. 166–7.

enlightened areas in the United States."[58] However, at the same time as the trial demonstrated the hostility towards Darwinism in the United States, a controversy was underway among Soviet Marxists over whether genetics or Lamarckism best illustrated the biological principles of dialectical materialism. Genetics was appreciated for providing fundamentally materialist rules of heredity, but also criticized for failing to—like Lamarckism—account for the role of the environment in evolution. Vavilov stayed aloof of Marxist discussions, and by the time of the collectivization drive during the first Five Year Plan, ideological debates on heredity and evolution were eclipsed by the need to raise yields and provide food.[59] As will be shown below, Lysenko's work was initially promoted based upon the criteria of practicality, not for reasons which had anything to do with whether or not acquired characteristics are inherited.

## 1.4 Julian Huxley and Leslie Clarence Dunn

The man who had invited H.J. Muller down to Texas was British biologist Julian Huxley. Though Huxley returned to England soon after the outbreak of World War I, he and Muller continued a life-long friendship begun during frequent walks on the arid, flat prairie outside Houston.[60] Huxley had a striking handsomeness that would age into an urbane air: He was known as a "statesman of science."

Julian Huxley was born amid the firework celebrations for Queen Victoria's Jubilee in 1887, and had an impressive intellectual pedigree. His grandfather, Thomas Henry Huxley had been the public defender of Darwin's theory of evolution—a dramatic figure known as "Darwin's Bulldog." After first reading *On the Origin of Species* T.H. Huxley said: "How stupid of me not to have thought of that."[61] Huxley was famous for having declared in a public debate at Oxford, with Bishop "Soapy Sam" Wilberforce in 1860, that he would rather be descended from apes than from a man who attacked a theory he was too ignorant to understand. Huxley also coined the term "agnostic" because he rejected the term "atheist": Neither the existence nor non-existence of God could be proved.[62] His grandson Julian's fascination with the question of nature vs. nurture could be attributed to

---

[58] Israel Agol, who would later become a close friend and associate of H.J. Muller's. See chapter 4. David Joravksy, "Soviet Biology and Marxism Before Lysenko," *Journal of the History of Ideas* 20, no. 1 (1959): 93.

[59] A useful summary of this period is provided by David Joravsky, "Soviet Biology and Marxism Before Lysenko," cited above. In addition see Joravsky's *Soviet Marxism and Natural Science, 1917–1932*, which is also cited above.

[60] Carlson, *Genes, Radiation, and Society*, p. 93.

[61] A recently published account of T.H. Huxley's reaction to Natural Selection and his role as Darwin's principle defender is Sherrie Lyons, *Thomas Henry Huxley: The Evolution of a Scientist* (Amherst, NY: Prometheus Books, 1999).

[62] Julian Huxley, *Memories* (Harmondsworth, England: Penguin Books, 1972), p. 9.

lifelong insecurity over whether his success was really his own, given the heroic reputation of his ancestor in scientific circles. Julian Huxley suffered repeated nervous breakdowns, for which he received electro-shock therapy, and almost refused knighthood due to a depressed sense that he did not deserve it.[63]

But Huxley was famous, with regular appearances on the BBC and lists of "Britain's Five Best Brains," compiled by *The Spectator*. He published widely, and had a talent for explaining complex ideas to a lay audience. The title of Huxley's most well-known work, *Evolution, The Modern Synthesis*, would later give scientists the term used to describe the use of Mendelian genetics to explain Darwin's theory of Natural Selection.[64] Huxley would also be forced, thanks to Lysenko, into becoming one of the most prominent defenders of this advance, to those without the training to understand it on their own.

Another biologist who was deeply interested in educating the public on the science of heredity and evolution, and whose career would be profoundly impacted by the Lysenko affair, was Leslie Clarence Dunn. Dunn was the fifth child born to a "land-poor" family in Buffalo, New York in 1893. In his oral history, recorded from 1958 to 1960, Dunn described his development as a geneticist in a culture influenced by both Christian fundamentalism and faith in eugenics. Two of his older siblings died in a diphtheria epidemic, and the family struggled to work some profit out of their farm.[65] His aunt became the first female ordained Methodist minister, and took him to camp meetings on two occasions. As Dunn later recalled: "It was a memorable experience; the frightening kind of hysterical performance that one saw at camp meeting rather conditioned me against it."[66] His father kept him home from church on Sundays to read Mark Twain—*Life on the Mississippi* and *The Gilded Age*—and died when Dunn was 9 years old.[67] Unlike Huxley, Dunn was raised in an environment where the influence of religion prevailed over science.

Dunn read T.H. Morgan's *Heredity and Sex* when he was 21 years old, and studying on a scholarship at Dartmouth.[68] Dunn applied to study with Morgan after graduation, but was turned away from the already over-crowded fly room. He went to Harvard instead where he worked on the genetics of mice under William E. Castle.[69] Dunn's graduate work was briefly interrupted when he left to fight in World War I.

[63] C. Kenneth Waters and Albert Van Helden, eds. *Julian Huxley: Biologist and Statesman of Science* (Houston, TX: Rice University Press, 1992), pp. 1–15.

[64] Julian Huxley, *Evolution, The Modern Synthesis* (New York and London: Harper and Brothers, 1943).

[65] The Reminiscences of L.C. Dunn, Oral History Research Office. Columbia University, 1961, pp. 1–3.

[66] The Reminiscences of L.C. Dunn, p. 7.

[67] The Reminiscences of L.C. Dunn, p. 11.

[68] The Reminiscences of L.C. Dunn, p. 21.

[69] The Reminiscences of L.C. Dunn, pp. 27–9, 47–9.

Dunn was stationed in France, and often arranged to travel through Paris so he could visit biology labs. Dunn was surprised by the primitivity of the work conducted, and later wondered why the French paid so little attention to genetics. It seemed to him paradoxical, given their passionate interest in evolution.[70] In 1918, after his return to the United States Dunn married Louise Porter. Dunn was charismatic, and many years later, after he appeared on a television talk show to discuss nature vs. nurture, a female production assistant marveled that he really ought to be in show business.[71]

Dunn had come across eugenics in college, in a book called *The Social Direction of Human Evolution* by William E. Kellicott. It seemed sensible, as he put it, because "one was aware at that time of a lot of idiocies and imbecilities where inheritance played some role, and one thought more, in those times, of what was then called the social burden of the unfit."[72] At Harvard Dunn became involved in a project to study the racial mixture of the Hawaiian Islands—including Japanese, Chinese, contract labor like Indians and "American Negroes"—to learn about the inheritance of traits by which races differ. Once the skeletal measurements were dropped on Dunn's desk it was his job to do something with the data, but it was very discouraging work.[73]

Dunn's distaste for eugenics would develop further following greater contact with eugenicists and their research. These early encounters likely influenced his later active engagement in educating the public about race and heredity. They also probably inclined him to be more sympathetic than he would otherwise have been to Lysenko's crusade against "biological determinism."

Unlike Dunn, Julian Huxley had a keen interest in eugenics, and believed it would ultimately be proven that "Negroes" were intellectually inferior.[74] In 1932 Huxley's brother Aldous published a dystopian account of a future governed by scientific totalitarianism. The book was inspired by his recent trip to the United States and ensuing disgust with everything American—from chewing gum, to moving pictures, to Henry Ford. The book was entitled *Brave New World.*

H.J. Muller rejected the pessimism of *Brave New World*. Like Huxley he believed eugenic measures—such as the establishment of sperm banks preserving the seed of the superior—could play an important role in creating a well-ordered society. In a public lecture he was asked to give by the president of Rice University in 1916, Muller warned his audience that in the past, Natural Selection meant that inferior members of the human race were eliminated by disease, war and famine. Modernity and the advancement of civilization however, meant those who would have died off were now supported and able to live. Even worse, the shiftless and less intelligent tend to breed at a higher rate. Muller described this as a crumbling of the quality of

---

[70] The Reminiscences of L.C. Dunn, pp. 136–9.

[71] Wolfe, "Speaking for Nature and Nation: Biologists as Public Intellectuals in Cold War Culture" pp. 72–3.

[72] The Reminiscences of L.C. Dunn, p. 165.

[73] The Reminiscences of L.C. Dunn, pp. 171–2.

[74] Julian Huxley, *Man in the Modern World* (New York: Mentor Books, 1944), p. 41.

the human race, though as a scientist he allowed it was not yet known what about us is actually passed on to our children. Some audience members later complained about Muller's lecture, not because of what he had said about eugenics, but because he had endorsed evolution.[75]

The Russian eugenics movement, led by Nikolai Koltzov and Iurii Filipchenko, was officially founded in 1920—well after eugenics movements were established in the West.[76] It differed from its counterparts in the U.S. and Western Europe in ways which are significant, and primarily related to issues specific to Soviet biology.[77] Filipchenko headed the Bureau of Eugenics in Petrograd, where funding also helped support his students in genetics, including Theodosius Dobzhansky. Like Koltzov, Filipchenko believed that eugenics could act as a "civic religion," promising a better future. However, eugenics also served the important purpose of making the bureau's work in genetics seem socially relevant.[78]

Throughout the 1920s Russian eugenics was dominated by two groups—the liberal intelligentsia and Bolshevik eugenicists. The former group provided scientific respectability while the latter gave the movement ideological legitimacy.[79] Soviet eugenicists were well aware of the policies being promoted in the West, however much of it seemed irrelevant since the situation in Russia after the Revolution was so severe. Immigration restrictions and sterilization laws made little sense in a country few seemed interested in immigrating to, and where the population was declining. Moreover, eugenicists like Filipchenko were appalled by the notion of government mandated sterilization, which Soviet eugenicists referred to as the "Indiana Idea."[80]

By the middle of the decade, however, support for eugenics became problematic for reasons beyond the controversial issues surrounding negative eugenics. Soviet eugenicists were also involved in the greater controversy over whether genetics or Lamarckism was the biological science most compatible with Marxist revolutionary dialectical materialism. Filipchenko defended genetics as Marxist science by pointing out that if Lamarckism were correct, the dysgenic conditions of poverty and oppression would have prevented the working class from ever being able to rise up against the bourgeoisie. Muller's 1927 breakthrough using X-rays to

---

[75] Carlson, *Genes, Radiation and Society*, pp. 103–5.

[76] Adams, *Wellborn Science*, pp. 155, 159–60.

[77] For a fascinating comparison of the eugenics movements in the Soviet Union and Nazi Germany see Loren Graham, *Between Science and Values* (New York: Columbia University Press, 1981), pp. 217–56.

[78] Ibid, pp. 163–4. With reference to the "utility of the imprimatur 'eugenics' for supporting 'genetics,'" Adams points out that when the Bureau of Eugenics was renamed the Bureau of Genetics and Eugenics it published no eugenics research (Adams, *Wellborn Science*, p. 164).

[79] Ibid, p. 171.

[80] Ibid, pp. 174–6; In 1907 Indiana became the first state in the U.S. to legalize involuntary sterilization.

produce mutations in *Drosophila* also gave genetics a boost by proving that heredity is not fixed, but can be altered by environmental influence.[81]

Though eugenics survived the dispute between genetics and Lamarckism by proving adaptable to any interpretation of heredity, it would almost completely disappear in the Soviet Union after Stalin declared the Great Break in 1929. An article published in the *Great Soviet Encyclopedia* in 1931 referred to eugenics as a "bourgeois doctrine," as the Soviet Union, in a development similar to the campaign against genetics which would be launched 17 years later, became the first country in the world to "ban" eugenics.[82] These two events were not unrelated given that the popularity of eugenics in the United States, its association with genetics, the failure of geneticists to criticize the abusive uses to which their research was put—not to mention the collaboration of German geneticists with the Nazi regime—would later allow Lysenko to portray the two as indistinguishable. Even worse, Muller's attempt to convince Stalin of his eugenic views would be a significant part of the reason why the VII International Congress of Genetics was not held in Moscow.

## 1.5 J.B.S. Haldane, and Dunn's Visit to the Soviet Union

In 1927—the year Dobzhansky left for the United States and the Soviet Union celebrated the tenth anniversary of the October Revolution—Leslie Clarence Dunn first visited. Dunn traveled to the USSR on behalf of the Rockefeller Foundation, to investigate the work of Soviet biologists and increase international exchange between scientists and researchers.[83] Dunn was enchanted by the Soviet Union, and this early encounter would color his views of the USSR for the rest of his life.

The emergence of foundations like Rockefeller was part of a transformation in how scientific research was funded in the United States in the early twentieth century.[84] This change was relatively concurrent with the founding of the Soviet Union, however conditions in Russia and its relationship with the United States delayed any kind of extensive engagement or contact until the early 1920s. Scientists like Dunn who arrived later in the decade encountered impoverished conditions, though the atmosphere of fear and anxiety Rockefeller representatives would witness a decade later—along with a significant improvement in terms of equipment and

[81] Adams, *Wellborn Science*, pp. 177, 179.

[82] Ibid, pp. 154–85.

[83] Melinda Gormley, "Geneticist L.C. Dunn: Politics, Activism and Community" (Ph.D. dissertation, Oregon State, 2006), 74–6.

[84] For an overview see Robert E. Kohler, *Partners in Science: Foundations and Natural Scientists, 1900–1945* (Chicago: University of Chicago Press, 1991); Nikolai Krementsov and Susan G. Solomon, "Giving and Taking Across Borders: The Rockefeller Foundation and Soviet Russia, 1919–1928," *Minerva* 3 (2001): 265–98.

organization—was largely absent.[85] Once the Cold War got underway the Rockefeller Foundation became increasingly aware of its own role in combating communism and the necessity of taking an applicant's political opinions and activities—including attitudes towards "leftist Lysenkoism"—into account.[86] For this reason the enthusiasm for Soviet science Dunn developed on his single visit to the USSR would prove problematic for his career.

Dunn had co-written a textbook, *Principles of Genetics*, with Edmund Sinnott, which was already in use in the Soviet Union when he arrived.[87] In the book Sinnott and Dunn strongly advocated the implementation of eugenic policies, and offered positive assessments of the Juke and Kallikak studies. While the authors admitted that studies of the inheritance of mental traits still did not account for the role of training and environmental influence, they also claimed that such evidence was "at hand."[88] Dunn was approached about a Russian translation and readily agreed.[89] However, the Russian edition omitted the chapter on eugenics, and said in the preface—"Students in the Soviet Union who have been using this book in the English edition will note that it follows the English very faithfully, except for the omission of the last chapter, which contains some outmoded bourgeois capitalist ideas which can have no relevance for any member of our body of citizens." Once Dunn discovered this, he told his co-author—"Well, maybe I'm one of those Reds too, but I don't like the last chapter any better than they did in the Soviet Union. Suppose we chuck it"—and the chapter was removed.[90]

It never occurred to Dunn he should inform his publisher, McGraw Hill, of any of this. Once they found out they were pretty upset, particularly since the Soviet edition was so awful: Illustrations were simply photographed and reproduced on newsprint. It was, as Dunn put it, "a mess." Nevertheless, the book went on to sell more copies in the Soviet Union than it did in the United States, though neither Sinnott, nor Dunn, nor McGraw Hill ever received any royalties.[91]

---

[85] Raymond B. Fosdick, *The Story of the Rockefeller Foundation* (New York: Harper and Brother Publishers, 1952), pp. 286–7.

[86] John Krige, *American Hegemony and the Postwar Reconstruction of Science in Europe* (Cambridge, MA: MIT Press, 2006), pp. 134–9, 142–3.

[87] Edmund Ware Sinnott and L.C. Dunn, *Principles of Genetics: An Elementary Text, with Problems* (New York: McGraw-Hill, 1925).

[88] Edmund Ware Sinnott and L.C. Dunn, *Principles of Genetics: An Elementary Text, with Problems* (New York: McGraw-Hill, 1925), pp. 388–9.

[89] The Reminiscences of L.C. Dunn, p. 243.

[90] The Reminiscences of L.C. Dunn, pp. 252–4. In the second edition of *Principles of Genetics*, published in 1932, the material on eugenics was removed, except for a paragraph at the beginning (p. 14 of the 1925 edition, p. 13 of the 1932 edition) where the authors provided a brief, positive description. This paragraph can also be found in the third edition, published in 1939, however the word "eugenics" was removed from the index. Any reference to eugenics was dropped by the time of the fourth edition, published in 1950. Dobzhansky was also listed as a co-author on the fourth edition.

[91] The Reminiscences of L.C. Dunn, pp. 254–5.

Dunn also met Vavilov in Moscow during his trip to the Soviet Union. Dunn later said Vavilov led the charmed life of a tightrope walker—"he got things done, and his voice was heard, and he was paid attention to."[92] Dunn's report to the Rockefeller Foundation was enthusiastic—"too enthusiastic"—he would later say. Dunn saw the Soviets as a great untapped source of talent, and believed Americans could benefit by visiting Russia. However, it was clear the Rockefeller Foundation did not share this view. As Dunn remembered: "I wasn't quite restrained enough, because I had found such interesting things there. I had been very much stimulated myself. They sent somebody else, who submitted a different report."[93]

The year after his trip to the Soviet Union Dunn was invited by Columbia to replace T.H. Morgan, and his second son, Stephen, was born.[94] Stephen, suffered from cerebral palsy. Caring for a son who would be considered by many to be "unfit," absorbed much of Dunn's time and attention.[95] His son's condition also probably transformed Dunn's mistrust of eugenics into resentment.

At the same time Dunn became devoted to educating the public on scientific issues. As early as 1929 the "What Is Going On This Week" section of the *New York Times* advertised "Professor L.C. Dunn of Columbia University" giving a lecture on "Recent Advances in Genetics" at the American Museum of Natural History, 77th street and Central Park West, 8:15 p.m. Dunn spoke out frequently against the distortion of genetics to rationalize various prejudices. He later wrote: "One of the best services that a biologist of long experience can render his fellow-citizens is to tell them in language which they can understand what he thinks of the human species."[96]

Dunn arrived at Columbia just in time for the Great Depression. As he remembered, students avoided graduation because there were no jobs to be had, and the department did what they could to support them.[97] At the same time, many scientists and academics noticed there was one country where the Depression not only did not seem to have had an impact, but the work of their colleagues was given tremendous support: The Soviet Union.[98]

---

[92] The Reminiscences of L.C. Dunn, pp. 684, 295.

[93] The Reminiscences of L.C. Dunn, pp. 308–10. Unfortunately Dunn's report was not saved by the Rockefeller Foundation, however a letter from the president of the General Education Board of the Rockefeller Foundation, Wickliffe Rose, to C.B. Hutchison, dated November 23, 1927 indicates that the foundation was "very glad to have definite information concerning conditions in Russia" as a result of Dunn's trip. Rose to Hutchison, November 23, 1927, folder 177, box 12, 2B 05 1.1, Rockefeller Foundation Archives, RAC.

[94] His first son, Robert Leslie, had been born in 1921.

[95] B: D65 pt. 2 Dobzhansky on Dunn. Dobzhansky, Theodosius. Réminiscences, Part II. The American Philosophical Society, pp. 464–6.

[96] "What is Going on This Week," *New York Times*, March 10, 1929; "Practical Biologist," *New York Times,* March 19, 1944.

[97] The Reminiscences of L.C. Dunn, pp. 402–3.

[98] In *Beyond the Laboratory*, Peter J. Kuznick chronicles how the attitudes of scientists towards the Soviet Union evolved during the late 1920s and early 1930s. See *Beyond the Laboratory* (Chicago, IL: University of Chicago Press, 1987), pp. 112–43.

In 1928, one year after Dunn's visit, renowned British geneticist J.B.S. Haldane also traveled to the USSR. Haldane had a large physique and unpredictable temperament. He moved like a charging bull, and seemed perpetually in a china shop, rather than Cambridge University, where he was an eminent professor. Haldane was brilliant in several subjects, and largely self-taught in genetics. His mathematical work showing how genes with even minor selective advantages could be incorporated into a general population was an important contribution to the evolutionary synthesis. Haldane's calculations showed the cumulative power of slight difference.[99]

Haldane and Huxley had first met as at Eton, and become close confidants as students at Oxford. Haldane, like Huxley, came from a family of scientists, and possessed a similar talent for summarizing scientific topics for a popular audience. Haldane also had strong views on eugenics, which tended to contradict his opinions on other topics. However contradictions never seemed to trouble Haldane.[100] In contrast to Huxley's cool urbanity, Haldane was pure scandal. He was said to be an example of how to lose friends and influence people.

One of Haldane's more eccentric habits was self-experimentation. Haldane drank ammonium chloride, calcium chloride, sodium chloride and hydrochloric acid.[101] He said the body should be regarded with the same respect as the starry sky but also—in the name of science—used up. He also helped his father develop the gas mask used by British troops during World War I, and became a strong advocate for poison gas—as more humane than other types of weapons—once the war was over.[102] Such opinions only added to his notoriety.

Haldane's propensity to offer himself as an object of experimentation has been attributed to his "penchant for melodrama."[103] The experiments he performed on others, however, were motivated by his desire to teach and be understood. Haldane once gave a lecture in the church hall of a small village, where complained that the War Office had chosen not to use a gas made by heating cayenne pepper because it was not lethal. It would have been effective anyway though, Haldane argued, and produced a lamp and a spoonful of pepper to demonstrate. He held the spoon over the lamp and within the seconds the entire audience was coughing and rubbing their eyes. "If that upsets you," Haldane admonished, "how would you like a deluge of poison gas from air fleet in real war?"[104]

Haldane's walrus mustache resembled Stalin's, which he explained by saying if he had to look like a dictator he would prefer a Russian one.[105] His above-mentioned first and only visit to the Soviet Union, at the invitation of Nikolai Vavilov, occurred

---

[99] Carlson, *Mendel's Legacy*, pp. 236–8.

[100] See Diane Paul, "Eugenics and the Left," *Journal of the History of Ideas* 45, no. 4 (1984): 567, 585.

[101] Ibid, pp. 61–2.

[102] See J.B.S. Haldane, *Callinicus: A Defense of Chemical Warfare* (New York: E.P. Dutton and Co., 1925).

[103] Ibid, p. 161.

[104] Ibid, p. 76.

[105] Ibid, p. 176.

shortly after Haldane was married. Vavilov arranged a lavish reception for Haldane and his wife at the Moscow Institute, with champagne and dancing. They were taken to Lenin's tomb in Red Square, the churches, and the State museum in the Kremlin. They paid a special visit to Vavilov's institute in Leningrad where they viewed the summer palaces of Catherine the Great and Nicholas II. Haldane's wife, Charlotte, remembered Vavilov had "dark twinkling Tatar eyes," and loved wine and women.[106] Like Dunn and Huxley, Haldane left the Soviet Union impressed by the role of the revolution in connecting science and society.[107]

## 1.6 The Rise of T.D. Lysenko, and Huxley's First Visit to the Soviet Union

In 1929 Stalin proclaimed the Great Break, and Lysenko's career was already underway. At a remote research station in the Caucasus, he was assigned the humble task of figuring out which crops would be most suitable for the frigid local conditions. Disdaining the laborious, systematic work required, Lysenko invented something called "vernalization"—the transformation of winter-habited plants into spring habit. When he presented his results to experts at a scientific convention he was mortified to learn it had already been done. And though vernalization became the basis for Lysenko's belief he could transform nature, it was actually just a technique that had never proved useful, and never would.[108]

This was the reality, but in the mythology of Michurinism it became the occasion when Lysenko ignored the supercilious sneers of theoreticians and experts. Why waste years in repeated experiments just to give them a chance to prove you wrong? Just to create a "little green monster" to be used against you?[109] He would not grant them the satisfaction. Lysenko was not concerned with the debates of theorists and academics. Lysenko argued his ideas were ignored because bourgeois scientists lacked practical experience and the insight of intuition.[110]

And fate was working in his favor. In 1929, Lysenko's father—Denis—conducted his own experiments with vernalization. He had sown winter wheat in the spring after having kept the seeds in a sack under snow, and obtained an exceptional yield. It is likely the sacks were under snow because Denis Lysenko was hiding them from grain collectors. The seeds had sprouted accidentally, and he planted them in the

[106] Charlotte Haldane, *Truth Will Out* (New York: Vanguard Press, 1950), pp. 41–2.

[107] Clark, p. 97. Haldane was not impressed, however, by Russian medical genetics. See Adams, "Last Judgment," p. 458.

[108] See Joravsky's account of vernalization in the *Lysenko Affair*, pp. 187–201; Also Soyfer, *Lysenko and the Tragedy of Soviet Science*, pp. 12–5.

[109] Joravsky, *Lysenko Affair*, p. 191.

[110] Ibid, pp. 56, 62

hope they had not been ruined. The massive losses, hunger and starvation of previous years helped make the results seem miraculous.[111]

It was luck, and just in time. Lysenko seized the opportunity to launch a campaign promoting the achievement of his belief—brief though it proved to be. An investigating committee organized by the Commissariat of Agriculture visited Denis Lysenko's farm and soon vernalization was widely imposed as official policy. Reports in *Pravda* insisted the results were so miraculous as to be "beyond immediate calculation." They followed a "high road of vast possibilities."[112]

"Time does not wait!" was a slogan of collectivization, and Lysenko moved on to his next idea at the manic pace that would characterize his career. His theory of the "phasic development of plants" was mystical and maddeningly unspecific. Though characterized by statements such as—"we do not yet know all the phases" and "plants pass through other phases we do not know of"—it became fundamental to Lysenkoism.[113] This was because it allowed him to most convincingly invoke the authoritative term Michurinism.

Phasic theory was based on the idea that plants went through various developmental stages, and after understanding these stages we can "destabilize" and "break" their heredity to transform them. If exposed to new conditions at the proper moment in their development, plants would acquire the necessary characteristics to survive. Thus any species could change, become something new, and thrive in any environment. It was the same as Michurin's idea that plants could be "taught" and transformed by grafting—attaching cuttings to more mature plants they could "learn" from. The flowering fauna produced by such experiments would form the landscape of Michurin's legendary garden. In a socialist realist account of "Soviet Michurin agrobiology," V. Safonov's *Land in Bloom*, the author wrote,

> At the gates the visitors were obliged to leave their baggage of accustomed conceptions and traditional knowledge as one leaves one's umbrella and galoshes in the hall.
>
> It seemed as though the very power of the frowning sky and of stern winter ceased at these gates.
>
> A motley crowd of hitherto unseen plants welcomed the visitors. The branches of apple and pear trees were barely able to carry the weight of enormous fruits. The winding stems of Far Eastern Actinidia clung to poles in the ground, but here they bore large, heavy, sweet, amber-coloured berries that smelt and tasted like pineapple. Peaches fraternized with apricots. In one year almonds threw out shoots seven feet long. What looked like bunches of grapes hung from the branches of a strange tree—a blend of the sweet and sour cherry. And next to it a capricious southerner—a grape vine—waved its tendrils with their scalloped leaves in the light breeze.[114]

The Soviet Union became increasingly isolated during the 1930s. The pressure of political campaigns, lost jobs, arrests and purges took its toll, and scientists

[111] Soyfer, *Lysenko and the Tragedy of Soviet Science*, pp. 17–20.

[112] Soyfer, *Lysenko and the Tragedy of Soviet Science*, pp. 18–9.

[113] Joravsky, *The Lysenko Affair*, pp. 208–9.

[114] V. Safonov, *Land in Bloom*, trans. J. Feinberg (Moscow, Russia: Foreign Languages Publishing House, 1951), pp. 156–7.

found themselves caught in an atmosphere of paranoia, where the meaning of their work changed completely. Scientific papers came to require extensive digressions on the significance of the topic at hand in terms of Marxism-Leninism. Scientists no longer criticized one another, they "unmasked" and "exposed" dangerous ideas. Work identified as theoretical opened the author up to accusations of being separated from reality. Foreign scientists were termed "toadies of the bourgeoisie" and "lackeys of world capital," and Soviet scientists who referred to them were considered "lily-fingered" and "servile to the West." Citations from Marxist classics and Party leaders became the best defense. A useful guide was published entitled *Marx, Engels, Lenin on Biology*, which proved so essential to the task of providing a quote in a pinch, it ran into several editions.[115] The no-longer obscure Michurin was propped up next to world famous figures like Ivan Pavlov as a great patriotic practitioner.[116]

Academic conferences were replaced by "public discussions" of scientific issues. The public discussion was an important ritual, scripted by the humiliating practices of criticism and self-criticism. They became the arena where power struggles could be played out, favoring the ideologically astute over the scientifically competent. Honorary Presidiums were elected, composed of Politburo members and bureaucrats. The proceedings were followed not just in scientific journals, but the daily press as well. The spectacle routinely ended with formulaic letters to Stalin.[117]

In the summer of 1931 Julian Huxley visited the Soviet Union as part of an official delegation and wrote a series of articles on his experience for *Nash's Magazine*, *Vanity Fair*, and the *Week-End Review*. He also published a book, *Scientist Among the Soviets*, wherein he referred to two mutually contradictory myths about the Soviet Union. In one, it was a pathologically corrupt, ignoble tyranny, bent on cruelty and preserving power. The masses were exploited and "bamboozled" by propaganda as art, religion, and finer feelings were trampled. Morality did not exist and any opposition was oppressed with violence. In the other, "which may be styled the Roseate Russian legend," the people were far better off than before the Revolution, the collective farm was at last going to make country-life as attractive as the town, and workers were provided with free parks, theaters, clubs, medical attention, and a "fortnight's vacation." To ensure safety and progress the occasional "fool" or "knave" was arrested, but on the whole enthusiasm of the worker exceeded even the optimism of the official plan.[118]

The truth, Huxley said, was more complex—a hybrid produced by transition between the medieval past and communist future, guided by the scientific attitude required by Marxism. Among the evidence of this was public education in museums, including the Darwin Museum and Museum of Evolution in Moscow. The latter contained live specimens, and demonstrations of scientific experiments

[115] *Marx, Engels, Lenin on Biology* (Moscow, Russia: Party Publisher, 1933).

[116] Krementsov, *Stalinist Science*, pp. 47–53.

[117] Krementsov, *Stalinist Science*, pp. 46–51.

[118] Julian Huxley, *A Scientist Among the Soviets* (London: Chatto and Windus, 1932), pp. 27–9.

for visitors. Huxley noted that the Museum spread its influence by sending an assistant to "Moscow's most popular park" in the summer. The museum assistant brought live samples of brainless frogs to show that they were still capable of movement, and thus dispelled such old-fashioned notions as that life is dependent upon a soul.[119]

By this time Dobzhansky had decided not to return to the Soviet Union. In early September, 1931, the same year as Huxley's visit, Dobzhansky and his wife went to Vancouver, British Columbia to fulfill the bureaucratic requirements for U.S. citizenship. Since they were permanent staff members with academic tenure at Cal Tech they easily received visas, however the immigration inspector decided not to allow them back into the United States because, he concluded, Dobzhansky had broken the law by accepting a teaching job when he only had a student visa. The Dobzhanskys were panicked, and their daughter later recalled that there were hints of a suicide pact if they were forced to return to the Soviet Union.[120] Dobzhansky sent telegrams to everyone he could think of pleading for help. Once T.H. Morgan learned of Dobzhansky's plight, he got in touch with the president of Cal Tech, Robert A. Millikan, who happened to be vacationing with Herbert Hoover aboard the Presidential yacht. Hoover telephoned the Secretary of Labor who reversed the decision and finally the Dobzhanskys became U.S. citizens.[121] Had Dobzhansky been sent back to the Soviet Union it is almost certain that he would not have survived.

***

What do you think is the significance of the quote from Theophrastus that you read at the start of this chapter? Why do you think Lamarckian theories of heredity have been so persistent in our culture? What about them is appealing?

In the first section you also read about the backgrounds of Trofim D. Lysenko and Theodosius Dobzhansky. How were they similar and how were they different? It is worth keeping these differences in mind when you read how Dobzhansky reacted to the VASKhNIL conference over two decades later at the start of the Cold War.

---

[119] Ibid, 76. On his trip Huxley was hosted by Nikolai Bukharin, which indicates the degree of importance Soviet authorities attached to his visit. In *Biology in German and Russia-USSR* Eduard I. Kolchinsky writes that Huxley got the term "modern synthesis," which he would later use to describe combination of genetics and Natural Selection to explain evolution, from Bukharin. See Eduard I. Kolchinsky, *Biology in Germany and Russia-USSR: Under Conditions of Social-Political Crises of the First Half of the XX Century* (St. Petersburg, Russia: Nestor-Historia Publishers, 2007), p. 419.

[120] Coe, "Theodosius Dobzhansky: A Family Story," p. 20. William B. Province, "The Origin of Dobzhansky's Genetics and the Origin of Species." in *Evolution of Theodosius Dobzhansky*, ed. Adams, 106.

[121] Barbara Land, *Evolution of a Scientist: The Two Worlds of Theodosius Dobzhansky* (New York: Thomas Y. Crowell Company, 1973), pp. 185–9.

The historical thread connecting V.I. Lenin, Luther Burbank, I.V. Michurin and Nikolai Vavilov is also important. What similarities and differences does this reveal—both practical and utopian—about the economic role of science in U.S. and Soviet culture?[122] Were capitalism and socialism really that different?

Next you read about Hermann J. Muller. Muller is yet another case of a young American—a scientist, artist or intellectual—who was attracted to the Soviet Union during the interwar period. Though hindsight makes this attraction seem naïve, it is worth thinking about the reasons for their fascination, and whether or not you might have felt the same. Why or why not?

Why is it, or is it not, important that both genetics and Lamarckism could be convincingly used to support Marxist philosophy? Is the victory of Lysenkoism in 1948 proof of the unscientific nature, ironically, of Marxism? Why or why not?

Next, we have the connection between Muller and Huxley. What evidence of Darwinism do you see in their relationship? Muller was among the most combative talents in the history of science. So why did he not fight with Huxley? Also, why do you think eugenics is an important part of this story?

And how does Dunn fit in? Muller, Huxley and Dunn all shared a similar level of jejune fascination with the USSR, but would end up at entirely opposite poles of opinion on the subject two decades later. Pay attention to the question of "why" as we continue, and then ask: Who was really "wiser?"

As for Dunn's visit to the Soviet Union, it is characteristic of Dunn that the basic business principle of informing your publisher you have O.K.'d a Russian translation of your book eluded him. Why is it important to this story that it was the Russian version that (supposedly) inspired him to chuck the eugenics section?

Haldane is a bit different. As you will see, he became notorious for being the only prominent geneticist to support Lysenko. What elements of his personality, as described above, might have caused this?

Haldane and Huxley are also interesting to think about in comparison with Muller and Huxley. Like Muller, Haldane was highly combative. Yet, once again, he did not fight with Huxley. Is the reason similar to why Muller did not fight with Huxley? Remember this later when you read about Haldane's debate on the BBC and the fallout from it. Was Huxley a bridge between the two most radical elements on either end of the controversy—Muller and Haldane? How can the value of Huxley's diplomatic skills be described in terms of Natural Selection?

And then we have Lysenko's rise to prominence in the Soviet Union. What was Lysenko describing when he referred to a "little green monster?" Also, why is it ironic that the seeds Denis Lysenko is hiding from grain collectors are cited in support of his son's vernalization experiments?

How may one think of Soviet slogans like "Time does not wait!" in terms of evolution and Natural Selection? And think about Safanov's description of

[122] For other examples of these similarities see Susan Buck-Morss, *Dreamworld and Catastrophe the Passing of Mass Utopia in East and West* (Cambridge, MA: MIT Press, 2002).

Michurin's Garden. Is it not similar to descriptions of the Garden of Eden? Why are gardens such symbolic places in so many cultures?

Why is it ironic that *Marx, Engels, Lenin on Biology* was a best-seller for the reasons that it was? What do you think of the fact that Huxley was impressed that the museum assistant from the Museum of Evolution in Moscow in Moscow brought live samples of brainless frogs to Moscow's "most popular park" to prove that souls were a myth? How do you think that kind of thing would have gone over had Huxley tried it in England? Should scientists have a right to do something like that even if some people find it offensive?

How do you think Dobzhansky's relationship with the land where he was born ("had Dobzhansky been sent back to the Soviet Union it is almost certain that he would not have survived") plays into this story?

# Chapter 2
# Between World War and Cold War

*Joseph Stalin Spoke of millet.*
*To Michurin's pupils he spoke of dung and dry wind.*
*And the Soviet peoples' great harvest leader*
*Called the millet an unmanageable child.*
*But she was not the accused*
*As they interrogated the moody daughter of the steppes.*
*In Lysenko's greenhouse in distant Moscow*
*She testified to what helps and what disturbs her.*

...

*Against Hitler man and grain fought*
*Side by side on the once barren steppe,*
*The liberating army pushed forward*
*And millet followed lovingly its pace.*
*Drive the war-wolf from the beautiful land*
*And let the field reach to the last horizon!*
*Grain the earth should bear*
*Peaceful and happy would be the world!*
*Death to the fascists!*
*Root out the weeds!*

—from "The Rearing of Millet," by Bertholt Brecht[1]

## 2.1 The Sixth International Genetics Congress

In 1930 two Soviet students funded by the Rockefeller Foundation, Israel Agol and Solomon Levit, joined H.J. Muller at his lab in Austin. Their descriptions of rational economic planning in the USSR contrasted sharply with the observable

[1] Bertholt Brecht, "The Rearing of Millet," trans. Robert C. Conard in collaboration with Ralph Ley, *New German Critique* 9, 1976, pp. 142–52.

W. deJong-Lambert, *The Cold War Politics of Genetic Research: An Introduction to the Lysenko Affair*, Archimedes 24, DOI 10.1007/978-94-007-2840-0_2,

consequences of the free-market crash in the United States. Agol and Levit further inspired Muller's enthusiasm for Russian science. However they also let him know that Soviet Lamarckists were challenging genetics, with resentment deepening on both sides.[2]

The Great Depression was reflected in Muller's emotions. His marriage was suffering from long hours in the lab, and Morgan had sponsored one of Muller's contemporaries from the "fly room," Alfred H. Sturtevant, for membership in the National Academy of Sciences. Muller felt he deserved it more.[3] On January 10, 1932 Muller wrote a note complaining of the "predatory operations of T.H. Morgan." He shoved it in his pocket, left his lab, headed into the wooded hills outside Austin, and swallowed a roll of sleeping pills.[4] He was found the next day—dazed but alive—sitting under a tree.[5]

A few weeks later Muller learned he had received a Guggenheim Fellowship to work at the Kaiser Wilhelm Institute for Brain Research in Berlin. In the meanwhile he had become involved with the underground National Student League, a leftist student organization that was illegal on campus. Muller helped the League's members print and distribute their newspaper, not knowing that they, and now he, were under investigation by the FBI. The FBI informed the university of Muller's involvement in subversive, communist activities, which put them in an awkward position since Muller was a prominent faculty member. When Muller informed university authorities of his Guggenheim they were relieved he was leaving.[6]

In August, 1932, before departing for Berlin, Muller attended the International Eugenics Congress in New York City, and the Sixth International Genetics Congress in Ithaca. Unlike the biologists of Morgan's generation, Muller felt it was important to speak out against the most dangerous arguments of the American eugenics movement. At the Eugenics Congress he shocked his audience by asserting that it was an unplanned economy, rather than the individuals who suffered from it, which was to blame for the dysgenic elements in American society. Not only did he take a stand against genetic destiny, he came out as a socialist.

---

[2] Kuznick, *Beyond the Laboratory*, p. 120.

[3] Carlson, *Genes, Radiation and Society*, pp. 63–4.

[4] Though both of Muller's biographers, Elof Axel Carlson and James Schwartz cite the letter as being in the Edgar J. Altenburg Papers in the Muller collection at the Lilly Library at Indiana University (Carlson, *Genes, Radiation and Society*, p. 174, ft. 14; Schwartz, *In Pursuit of the Gene*, p. 249, n. 20), the letter was not, in fact, located there when I conducted research in the Lilly April 13–16, 2009. Those who wish to view the letter are advised to contact Sue Presnell or Cherry Dunham Williams at the Lilly for further information.

[5] A memo in Muller's FBI file contains testimony from a source whose identity is not revealed, saying "that about 1930 or 1932, Dr. Muller became insane, he believed temporarily so from an overdose of some drug." Federal Bureau of Investigation. Form No. 1. File no. 61-219. 8/19/41. Subject: Muller, Hermann J. Freedom of Information/Privacy Acts No. 1024867-000. Federal Bureau of Investigation. U.S. Department of Justice. Washington, D.C. 20535.

[6] Carlson, *Genes, Radiation, and Society*, pp. 173–8.

Muller's speech was given great publicity in the press. The *New York Times* quoted him saying that, "slums in our cities constitute veritable factories for the production of criminality among those who happen to be born in them, whether their parents were of the criminal class or not … Under these circumstances it is society, not the individual, which is the real criminal, and which stands to be judged." Even more controversial was the fact that he questioned whether success in a capitalist society was evidence of genetic worth: "The 'respectable' captain of industry, the military leader, or politician, and the successful gangster are psychologically not so far apart."[7]

News of Muller's speech, in the newspapers and on the national wire services, followed him from New York to Ithaca. Muller's paper at the genetics congress was complex, but made even more so by the nervous, disjointed way he presented it. According to Muller's authorized biographer, Elof Axel Carlson, his notes were jumbled between paper scraps and the backs of envelopes. Muller covered awkward silences while he searched for a sentence with non-sequitur asides. Morgan sat in the audience, struggling in vain to follow it, and commented to one of Muller's students, Bentley Glass: "Something is wrong with Muller."[8]

Leslie Clarence Dunn helped organize the Sixth International Genetics Congress in Ithaca. As secretary he was in charge of making sure delegates were welcomed and entertained. Nikolai Vavilov was the only Russian geneticist allowed to attend, and he was accompanied by Vladimir Saenko, the head of the agricultural section of the Soviet trade agency in the United States, who never left his side. Dunn drove down to the dock to meet Vavilov when he arrived, and Vavilov introduced Saenko as a "sort of potato breeder" and interpreter. It struck Dunn as odd that Vavilov would bring a translator since his English was "quite adequate."

"Now this meant something," Dunn later remembered, "The tightrope was getting a little slack. … This turned out to be his shadow."[9]

Dunn brought Vavilov and Saenko up to the Columbia University campus where foreign delegates were staying on the top floor of John Jay Hall. Dunn said to Vavilov, "Well, I was intending to take you home."

"Maybe it could be arranged," Vavilov replied softly, and Dunn understood.

The elevator stopped at the twelfth floor.

"Professor Vavilov this is your place," Dunn announced.

Saenko took his bag and got out. Dunn pushed the button and the elevator went down with Vavilov in it.[10]

---

[7] "Holds Capitalism Bars Eugenic Goal," *New York Times*, August 24, 1932.

[8] Carlson, *Genes, Radiation and Society*, p. 182; For a completely different account of Muller's presentation, see Schwartz, *In Pursuit of the Gene*, p. 253.

[9] The Reminiscences of L.C. Dunn, 686. In Dunn's Oral History he just says that Vavilov was accompanied by someone he described as an "interpreter" and "potato breeder." See footnote 15 below.

[10] The Reminiscences of L.C. Dunn, pp. 686–7.

Dunn brought Vavilov up to his house in Riverdale, a bucolic suburb in the Bronx, just outside of Manhattan. Three other geneticists were also staying with him, including J.B.S. Haldane. That year Haldane had moved from Cambridge to University College, London, and also published a book of essays in which he wrote: "The test of devotion of the Union of Soviet Socialist Republics to science will, I think, come when the accumulation of the results of human genetics, demonstrating what I believe to be the fact of innate human inequality, becomes important."[11] This quote would later come back to haunt Haldane once he emerged as Lysenko's most prominent defender.

Everyone was anxious to know why Vavilov was alone, but as Dunn later remembered, he "never took down his hair entirely."[12] Vavilov just said that he was on his way down to Peru after the conference because he was interested in maize and potatoes. It made more sense to bring an interpreter who knew something, than one who had no experience with agriculture at all. Dunn did not want to press him, and decided to just to consider it a nice evening among friends. Once the conference began Vavilov was accompanied constantly by Saenko, so no one had much chance to talk with him alone.[13]

Dobzhansky and Vavilov also met again at the congress. It had been two years since they had last seen each other in California, and they greeted one another in the usual Russian manner, with an embrace and a kiss. Speaking publicly Vavilov declared that Lysenko's techniques would make it possible to grow alligator pears and bananas in New York, and lemons in New England.[14] Dobzhansky was unable to speak with Vavilov privately until they arrived for lunch in a crowded cafeteria and found only two seats available at a table, forcing Saenko to find a place elsewhere. They conversed in Russian, which no one else at the table knew, but Vavilov still chose his words carefully. Dobzhansky understood from him that things in the Soviet Union had changed a great deal, and his own situation was no longer as it had been. This time Vavilov advised: "Dobzhansky, do what you want. If you want to return, do so. If you do not want to return, don't. Stay here."[15]

[11] J.B.S. Haldane, *The Inequality of Man and Other Essays* (London: Chatto and Windus, 1932), p. 137.

[12] The Reminiscences of L.C. Dunn, p. 686.

[13] The Reminiscences of L.C. Dunn, pp. 686–7.

[14] "Symbol of Stalinist Science," *New York Times*, November 24, 1976.

[15] The Reminiscences of Theodosius Dobzhansky, pp. 166–7. In his oral history Dobzhansky describes Vavilov has having been accompanied by two men from Amtorg, the agricultural trade agency in the U.S.—"who were officially his helpers, helping him to put up charts and that kind of thing. But these two individuals never left Vavilov alone. It was perfectly clear that they were far more than they were officially calling themselves." Nikolai Krementsov, however, informed me that Vavilov was actually accompanied only by a man named Saenko, the head of Amtorg, who was supposed to assess the exhibition, which was part of the genetics congress, for new varieties of crops etc., with the view of purchasing some of them for use in Soviet agriculture. Written Communication. Nikolai Krementsov. March, 2006. For further details on the congress see Krementsov, *International Science Between the Wars*, pp. 36–42.

After the congress Vavilov left for Peru. He and Dobzhansky never saw each other again, but Dunn met Vavilov when he returned to the United States on his way back to the Soviet Union. Vavilov had said he would be back in January, but it was already mid-February when he arrived in New York. Vavilov was staying at a hotel and Saenko was still with him. Dunn called to ask if Vavilov could come stay up in Riverdale. Vavilov sounded nervous over the phone, but agreed to at least have dinner. He told Dunn there were things he would like to talk to him about, but it would be better to do it face to face. It was close to dinner time when Vavilov arrived and they went immediately up to Dunn's study to talk: "The real reason why I'm late *is* because I collected—and don't let anybody else tell you any differently," Vavilov said.[16]

Vavilov returned to Russia and Dunn heard nothing further until one day a manuscript arrived, written by Lysenko, forwarded by Vavilov.[17] There was no comment aside from, "I've been handed this, asked to send it to an American journal." The paper, according to Dunn, was nonsense—completely lacking in proof or any kind of sustained analysis. Dunn sent it around to some other people who had the same impression. As he said—"We rejected the paper, with cause… One interesting point would be to find out whether he took umbrage against the people who had sent his manuscript out. He didn't send it under his own cover. … Now I believe that was the only manuscript of his that was offered for publication outside the Soviet Union."[18]

## 2.2 Muller Moves to the Soviet Union

Morgan had been right when he commented that something was wrong with Muller in Ithaca. Two months later Muller left the country alone. He and his wife Jessie understood it as a trial separation. They would later divorce. Though Muller did not know it at the time, it would be eight years before he would return to the United States. Vavilov had told Muller he would have all his expenses paid if he visited the Soviet Union. On the boat over Muller immersed himself in Lenin's major work, *Materialism and Empirio-Criticism*, and read other philosophers to prepare himself for the advanced writings of Marx and Engels.

One of Muller's colleagues in Berlin was a Soviet biologist, Nikolai Timoféef-Ressovsky, who had followed in Muller's footsteps using radiation with *Drosophila*. Timoféef-Ressovsky was anxious to collaborate with Muller and believed that biologists and physicists should work together to study the impact of radiation on the gene. During this period however Muller witnessed Hitler's rise

[16] The Reminiscences of L.C. Dunn, pp. 694–5.

[17] It's possible the manuscript was forwarded by another geneticist, Nikolai Koltzov. In his oral interview Dunn said that it was forwarded either by Vavilov, or Koltzov, but the detail emerges while he recounts his memories of Vavilov. The Reminiscences of L.C. Dunn, pp. 694–7.

[18] The Reminiscences of L.C. Dunn, pp. 696–7.

to power. Initially he did not take Hitler seriously, but over the course of the next year the Reichstag burned and all political parties but the Nazis were suppressed. The institute where Muller worked was ransacked and many of the scientific workers were beaten up.[19]

Vavilov stopped to visit Muller in Berlin on his way back from the Ithaca congress and his research trip to Peru, and offered him a position in Leningrad. For Muller it was an easy decision, given the worsening situation where he was. The Soviet Union promised full financial support and the backing of the most famous and powerful man in biology—or so it still seemed at the time.

Muller left for Leningrad with a suitcase full of flies and sent thousands of vials and bottles ahead so he could begin work immediately. Raissa Berg, a Russian geneticist assigned to his lab, would later remember she arrived to meet him "more dead than alive, overwrought at the prospect of seeing the great discoverer of the laws of nature."[20] The Russian Academy of Sciences elected Muller a corresponding member and granted him the freedom of a full-time research position. Muller's speech at the International Eugenics Congress in New York was translated and reprinted widely across the Soviet Union. He published articles in popular magazines praising the collective farms and support for science he saw around him, contrasting it with what he had left behind in the United States. Back home academics were losing their jobs and those who kept them saw their salaries almost cut in half.

Muller filled his time visiting genetics laboratories and delivering speeches. The Academy offered Muller his own well-staffed lab where he was joined by Agol and Levit, his students from Texas. The term "Morganism" was beginning to be attached to genetics to belittle it in the Soviet Union, particularly once T.H. Morgan won the Nobel Prize in 1933. Muller sought to disassociate genetics from Morgan—minimizing his influence and implying that Morgan had benefited mostly from the work of his graduate students—like Muller. He also portrayed genetics as Marxist, and criticized the still-popular Lamarckist approach to the study of evolution. Soviet Lamarckists referred to Engels' thesis on the importance of the transmission of characteristics acquired through labor in the transformation of ape into man.[21] As in the West, belief that acquired characteristics could be inherited remained very much alive.

To make genetics acceptable Muller described the fly room in terms which would appeal. Many of the members, he said, had come from working-class backgrounds, and the viewpoints they developed collectively, as a group, forced Morgan from his dictatorial position into recognizing their value. Muller believed that the Soviet Union offered an ideal testing ground for eugenic principles because everyone was

[19] Carlson, *Genes, Radiation and Society*, pp. 187–90.

[20] Raissa Berg, *Acquired Traits: Memoirs of a Geneticist From the Soviet Union* (New York: Viking, 1988), p. 28.

[21] Medvedev, *Rise and Fall of T.D. Lysenko*, p. 8.

equal. In the United States genetic worth was impossible to discern—people were born into privilege and there was no way of knowing if someone's success was really their own.

Muller's research team moved to Moscow in 1934. It was the beginning of unhappy change. Until then the Soviet Union had been second only to the United States in publications in genetics.[22] But in 1935 Vavilov was removed from his position as President of the Lenin All-Union Academy of Agricultural Sciences, Lysenko was appointed a member, and Michurin died.[23] Lysenko snapped up Michurin's legacy, perhaps understanding intuitively the power of those who are no longer around to be criticized. Though in reality Michurin had always avoided Lysenko, that was now the past. In the upheaval of the present they would be indistinguishable, as Lysenko dubbed his ideas "Michurinism."

Thousands of collective farms had been forced to adopt Lysenko's methods, and he swept away any suspicion they might not work with accusations of sabotage. In the Donetz Basin a miner, Alexei Stakhanov, revolted against—"old standards of output, the formidable 'technical norms,' sanctified by antediluvian textbooks, fossilized professors, and timorous specialists"—by drilling 102 tons of coal in 5 h and 45 min. The Stakhanovite revolt spread, and soon crankshafts were being stamped, automatic looms tended, and shoes stitched at record rates.[24] Brigades of Shock Collective Farmers were organized to lead the socialist advance on the village, and at their Second Congress in 1935 Lysenko declared:

> Tell me, comrades, was there not a class struggle on the vernalization front? In the collective farms there were kulaks and their abettors who kept whispering (and they were not the only ones, every class enemy did) into the peasant's ears: "Don't soak the seeds. It will ruin them." This is the way it was, such were the whispers, such were the kulak and saboteur deceptions, when, instead of helping collective farmers, they did their destructive business, both in the scientific world and out of it; a class enemy is always an enemy whether he is a scientist or not.[25]

Stalin was present and he cheered: "Bravo, comrade Lysenko, bravo!"

In January, 1936, a public letter to Stalin appeared on the front page of *Pravda* from Lysenko's parents entitled, "Could We Ever Have Dreamed of Such a Great Honor?" It read:

> Our beloved, dear Stalin! The day we learned that our Trofim was awarded the Order of Lenin was the most joyous day of our lives. How could we ever have dreamed of such a great honor, we, poor peasants from the village of Karlovka in the Kharkov province?

---

[22] Correspondence, L.C. Dunn to the *Saturday Review of Literature*. B: D917 L.C. Dunn Papers. Lysenko Controversy in the U.S. #2. American Philosophical Society.

[23] Carlson, *Genes, Radiation and Society*, pp. 205–6.

[24] Joshua Kunitz, *Along Came Stakhanov* (Moscow, Russia: Co-Operative Publishing Society of Foreign Workers in the USSR, 1936), pp. 1, 30; See also Mary Buckley, *Mobilizing Soviet Peasants: Heroines and Heroes of Stalin's Fields* (Lanham, MD: Rowan and Littlefield Publishers, 2006).

[25] Medvedev, *Rise and Fall of T.D. Lysenko*, p. 17.

> It was hard for our son Trofim to get an education before the Revolution. He was not admitted—a peasant boy, a muzhik's son—into the agronomy school, even though he received only the highest grades in school. Trofim had to become a gardener in Poltava. He would have remained a gardener for life had it not been for the Soviet regime. Not only the older Trofim, but his younger brother and sister went to study at institutes… The high road of knowledge was opened up to the muzhik's son… Is there any other country in the world where the son of a poor peasant could become an academician? No! …
>
> We do not know how to show you our gratitude, dear Comrade Stalin, for this great happiness, the conferring of the highest award upon our eldest son. … for work is enjoyable now, for life has become better and more cheerful. …[26]

Vavilov had initially placated Lysenko, believing Soviet science was a realm large enough to accommodate all points of view. He said Lysenko was an "angry species" who "walked by faith, not by sight," and that "all progress in the world had been made by angry men." Who knows, Lysenko might even "discover a way to grow bananas in Moscow?"[27] But now Lysenko and his followers accused Vavilov of inexcusable delays. When asked what genetics had to offer to increase agricultural productivity, Vavilov answered practically, referring to the time required to select better varieties.[28] Lysenko meanwhile rattled off promises of utopia, the transformation of nature and the elimination of enemies. Vavilov made the mistake of appointing Lysenko to the organizing committee of the 1937 International Genetics Congress to be held in Moscow. The Congress was cancelled and a year later Lysenko became president of VASKhNIL: The method of progress by angry men.[29]

Lysenko's presence on the organizing committee was not the only reason the congress was cancelled. In May, 1936, against his colleagues' advice, Muller sent Stalin a copy of *Out of the Night* with a letter explaining his ideas on eugenics. "As a scientist with confidence in the ultimate Bolshevik triumph throughout all possible spheres of human endeavor," Muller wrote, he had decided to address the matter directly to Stalin.[30] Trusting in Stalin's "farsighted view" and "strength in the realistic use of dialectical thought," Muller explained that artificial insemination using the reproductive material of the most superior—the one in 50,000, the one in 100,000—could ensure the triumph of socialism. It only takes a few years before a child can be recognized as backward or superior. After 20 years results would be noticeable, and if capitalism still existed beyond Soviet borders, "this vital wealth in our youthful cadres, already strong through social and environmental means, but then supplemented even by the means of genetics, could not fail to be of very considerable advantage to our side."[31]

---

[26] *Pravda*, January 3, 1936. Also available online at http://www.cyberussr.com/rus/lyse-par.html

[27] These memories were recounted by S.C. Harland in his debate with J.B.S. Haldane on the BBC and published in "The Lysenko Controversy: Four Scientists Give Their Points of View," *The Listener*, December 9, 1948, p. 873.

[28] Carlson, *Genes, Radiation and Society*, p. 225.

[29] Krementsov, *Stalinist Science*, p. 61.

[30] Mark Adams, "Eugenics in Russia, 1900–1940," in *The Wellborn Science*, ed. Adams, p. 195.

[31] Ibid.

Now Muller learned that Stalin was not happy with what he had read and was ordering the book to be attacked. All reviews to be published in the press were brought to a halt, awaiting further word. The secretary who had translated the manuscript into Russian was later arrested and shot.[32]

The Great Terror began in August. The lists of victims, arrests and executions absorbed most of the important allies of the geneticists. The Terror was too unwieldy an instrument to assign motivation, purpose or intent—too nonsensical to be able to say Lysenko had a line to the secret police and had all of his enemies eliminated. Some of his followers died too. But the ratio of repression is clear, and soon Vavilov would be gone as well.[33]

In December, 1936, a public discussion was held on "issues of genetics," with Muller, Vavilov and their colleagues on the defensive, and Lysenko and his associates ready to push further. They were watched by an audience of 3,000, composed of academics and representatives of the collective farms. They attacked one another personally, cheered their allies, hissed their opponents, and made accusations of fascism and anti-Darwinism. That same day it was announced that Agol had been arrested as an "enemy of the people."[34]

Lysenko said geneticists only imagined they saw genes in their microscopes. They viewed details of the cell—the nucleus, other bits—but not what they thought they were seeing. Muller countered that choosing between genetics and Lamarckism was like choosing between astronomy, chemistry and medicine, or astrology, alchemy and shamanism. He also revived Iurii Filipchenko's argument by pointing out that if Lamarckism were correct, then the Bolshevik Revolution could never have happened: poor living conditions would doom peasants and workers to a condition of perpetual inferiority. Muller's speech was met with wild applause, but his remarks were omitted from the published record.[35]

Muller was disillusioned and decided it would be best to go away for a while. He did not want to depart in any kind of obvious way, as though he had changed his mind about anything. The Spanish Civil War had broken out that year and an illustrious list of communists, romantics and adventurers (including Ernest Hemingway, André Malraux, George Orwell, W.H. Auden, John Dos Passos, Antoine de Saint-Exupéry, and J.B.S. Haldane) had gone to fight Francisco Franco and the fascists. Muller joined them for a number of months and then returned to the Soviet Union to pack his bags for good.[36]

---

[32] Adams, *The Wellborn Science*, pp. 196–7; See also Schwartz, *In Pursuit of the Gene*, pp. 263–8.

[33] Krementsov, *Stalinist Science*, pp. 57, 61. Joravsky, *The Lysenko Affair*, p. 118. Joravsky writes that the number of "arrested geneticists, philosophical defenders of genetics" and "all the other non-Lysenkoite biologists and agricultural specialists" adds up to seventy-seven. Only six of Lysenko followers were arrested. Joravsky, *The Lysenko Affair*, p. 116.

[34] Carlson, *Genes, Radiation and Society*, p. 233; Adams, *The Wellborn Science*, pp. 196–7.

[35] Joravsky, *The Lysenko Affair*, pp. 103–4; Soyfer, *Lysenko and the Tragedy of Soviet Science*, pp. 82–9; Krementsov, *Stalinist Science*, pp. 59–60.

[36] Carlson, Genes, *Radiation and Society*, pp. 235–43; Kuznick, *Beyond the Laboratory*, p. 137.

Muller's last night in Leningrad he saw the film *Peter the First* with Vavilov and Raissa Berg, then walked around the city until after midnight. Vavilov was completely refreshed a few hours later when he picked Muller up at his hotel, and took him on a farewell tour of his plant-breeding institute. The garden contained samples from all over the world and a laboratory testing the qualities of cereal cultures produced tiny loaves of bread that looked like communion wafers. They breakfasted on bread, tea, smoked fish and chocolate, and then returned to Leningrad. Muller left Vavilov an official letter to explain his departure. Muller wrote he would be back in two years, then crossed it out and wrote one. The day Muller left Agol was shot and Levit suffered a similar fate a few years later.[37] Muller never returned to the Soviet Union and he never saw Vavilov again.[38]

## 2.3 "Doby" and "Dunny"

In April, 1936, eight months before the public discussion on genetics in Moscow, Dobzhansky was invited by Leslie Clarence Dunn to give a series of lectures at Columbia under the title *Genetics and the Origin of Species*.[39] The title was bold but necessary, as what Dobzhansky discussed was the dramatic new direction the study of evolution had taken. He and his wife rented an apartment on 110th street near Columbus Avenue, where at the time the subway still ran above ground. Their apartment was unpleasant, boasting little more than a view of the elevated tracks.

Once the lectures were over Dobzhansky returned to California to turn them into a book. Work was assisted by a riding accident. A horse, blind in one eye, smashed Dobzhansky's left knee into 14 pieces against a cement post. At the operation the anesthesiologist believed Dobzhansky was drunk because he could not put him to sleep, unaware of the high tolerance for ether Dobzhansky had developed from his work with *Drosophila*. Bedridden for three weeks, Dobzhansky spent morning till evening writing the book that would make him among the most famous evolutionary biologists in the world.

In the preface to *Genetics and the Origin of Species* Dunn described the significance of Dobzhansky's work. The science of genetics was created because a rigorous method was required to study something that had originally been studied for its own sake—variation and heredity. The need for tests and experiments drove genetics into the lab and created a culture defined by bottles of *Drosophila*. To other biologists

[37] Adams, *The Wellborn Science*, pp. 196–7.

[38] Berg, *Acquired Traits*, pp. 38–40.

[39] Correspondence, Leslie Clarence Dunn to Theodosius Dobzhansky, April 17, 1936. B: D65 Theodosius Dobzhansky Papers. Leslie Clarence Dunn #3. The American Philosophical Society.

the work of geneticists seemed narrow: "…at its end they thought they could see a red light and sign 'The Gene: Dead End'."[40] But Dobzhansky's book was a sign that genetics was now ready to emerge and be tested in the "ultimate laboratory of biology, free nature itself."[41] He described the solution to the nature vs. nurture debate: Characteristics which exist are not created in direct response to the conditions of the outer environment, but rather due to random genetic mutation. However the environment plays a vital role in selecting which features survive. This was the "modern" or "evolutionary" synthesis of genetics and Natural Selection.

Afterward Dobzhansky was invited to teach at Columbia where he and Dunn became close. The relationship between "Doby" and "Dunny," as they were known to friends, grew into a collaboration that would include working to undermine Lysenko. However the outcome for Dunn was quite different than Dobzhansky. Dunn's views never submitted to the simple division defined by the Cold War, and he cared too much about politics to behave politically. Years later in Dunn's obituary Dobzhansky wrote, "to those who knew him well he was always the same 'Dunny,' generous and idealistic to a fault."[42]

In December, 1936, the same month Muller debated Lysenko in the wake of the cancelled International Genetics Congress in Moscow, Dobzhansky wrote to Dunn:

> Gogol said in one of his stories that whenever the devil touches gold it becomes chips of pottery. This seems to be true in USSR, and the fate of the genetics congress is a case in point.

He noted that though Lysenko was an "old moron and madman at the same time," the leaders of the Soviet Union "are not idiots." However they had "taken council of an idiot" and Dobzhansky suggested that a group of American geneticists send a treatise to Moscow, written in "*their*" language, explaining how genetics was not a Nazi theory: "Muller sits there and he is probably saying it all the time, but probably many peoples there no longer believe him."[43] A few weeks later Dobzhansky, still unsure of the situation in Soviet biology and forced to rely on the press for information, wrote to Dunn again.

> The recent information in New York Times suggests that the situation in Russia is not so bad as it seemed at first, and that it may finally regulate itself, as far as Russian genetics is concerned. It seems to me that for the time being it is best not to try to do anything more from this side, and just wait for developments.[44]

Dunn responded that he was amused to see how closely their ideas coincided, and agreed something describing how genetics was "not counter-revolutionary,"

---

[40] Theodosius Dobzhansky, *Genetics and the Origin of Species* (New York: Columbia University Press, 1937), p. viii. Dunn's "Editor's Preface" was dropped from later editions.

[41] Ibid.

[42] B: D65 Dobzhansky Papers. Dunn, Leslie Clarence #1. The American Philosophical Society.

[43] Correspondence, Th. Dobzhansky to L.D. Dunn, December 21, 1936. B: D917 L.C. Dunn Papers. Dobzhansky, Theodosius 1936–1937. The American Philosophical Society.

[44] Correspondence, Th. Dobzhansky to L.C. Dunn, January 4, 1937. B: D917 L.C. Dunn Papers. Dobzhansky, Theodosius 1936–1937. The American Philosophical Society.

"written in the Marxist dialect," might be just the thing.[45] In a follow-up letter Dobzhansky asked if Dunn had also heard that Muller had been arrested and was being held in a "socialist jail (which of course exists in order to re-educate the 'criminal' rather than to punish him)."[46] Dunn responded with news of Muller he had recently received from a journalist who was in contact with him. It seemed that, "one of Muller's men," Levit, had been forced to publicly declare himself a "fascist pig." Muller himself had been "roundly balled out," and was now apparently in a "terrific emotional stew," and "willing to go to any foolish extreme to square himself" with Soviet authorities. Despite the turmoil, however, Dunn assured Dobzhansky that Muller was not in prison, but had only temporarily left the Soviet Union to serve in the Spanish Civil War.[47]

The American Genetics Society appointed a three-member committee, which included Dunn, to develop a response to events in the Soviet Union. Dunn urged moderation by pointing out: "This is just as difficult for us to recognize as it was for the Europeans to understand our Monkey Trial in Tennessee. The two cases, being based on heresy hunts, seem quite analogous to me."[48]

Dunn's devotion to educating the public on scientific matters was by now matched with an intense enthusiasm for political causes. In 1933 Dunn had served on the executive committee of the Emergency Committee in Aid of Displaced German Scholars, organized to provide positions for intellectual exiles from Nazism.[49] Once the Spanish Civil War broke out Dunn, like Muller, became focused on the confrontation between communism and fascism, playing out like a rehearsal for a later, greater conflict. Muller volunteered as a way to leave the Soviet Union at an opportune moment, while Dunn engaged it in his own usual way—signing a letter. The Catholic Church came out against Spanish communists and there was Dunn's name in the *New York Times* beneath the headline, "Open Letter in Reply to Spanish Hierarchy's Recent Views of War."[50] Protests broke out in New York City in response Nazi atrocities against Jews; as thousands gathered in Columbus Circle to hear speeches by Orson Wells and Mayor LaGuardia, Dunn signed an open letter to President Roosevelt suggesting the severing of trade relations.[51] His concern for Spain led Dunn to become vice-chairman of the University Federation for Democracy

---

[45] Correspondence, L.C. Dunn to Th. Dobzhansky, January 7, 1937. B: D917 L.C. Dunn Papers. Dobzhansky, Theodosius 1936–1937. The American Philosophical Society.

[46] Correspondence, Th. Dobzhansky to L.C. Dunn, March 10, 1937. B: D917 L.C. Dunn Papers. Dobzhansky, Theodosius 1936–1937. The American Philosophical Society.

[47] Correspondence, Ronald Lund to L.C. Dunn, March 19, 1937. Correspondence, L.C. Dunn to Th. Dobzhansky, March 18, 1937. B: D917 L.C. Dunn Papers. Dobzhansky, Theodosius 1936–1937. The American Philosophical Society.

[48] Kuznick, *Beyond the Laboratory*, pp. 134–5.

[49] Gormley, *Geneticist L.C. Dunn: Politics, Activism and Community*, p. 2.

[50] "Open Letter in Reply to Spanish Hierarchy's Recent Views on War," *New York Times*, October 4, 1937.

[51] "Catholics of U.S. Score 'Atrocities'," *New York Times*, November 17, 1938.

and Intellectual Freedom in 1937. The following year he joined the Lincoln's Birthday Committee for Democracy and Intellectual Freedom, organized to educate the public on the meaning of the term "race."[52] Dunn's identities as a scientist and as a political activist would soon prove difficult to distinguish, or to reconcile.

Press reports in the West concerning the cancellation of the International Genetics Congress in the Soviet Union quoted Lysenko saying genetics was just a game—like chess or football.[53] The association between genetics and eugenics was also working in Lysenko's favor. An article which appeared both in *Science* and the *New York Times* quoted a "party representative" accusing a member of Levit's staff, Professor Shtyvko, of "making deductions 'resembling the racial nonsense of German fascists'." The charges referred to an article Shtyvko had published in a German scientific journal where, after studying the skeletons of 54 starvation victims, he concluded that the strains of famine and civil war had reduced them to being "somewhere between the Germans and the yellow race." In another paper Shtyvko had determined that the Buryat Mongols of Siberia were the mental equivalent of European 12 year olds. His work was cited as proof that Soviet science must be protected from any "anti-scientific theories that might be dragged in."[54] Several months later the relationship between Nazi eugenics and the eugenics movement in the United States was reinforced when the University of Heidelberg awarded an honorary degree to Henry Laughlin from the Eugenics Record Office in Cold Spring Harbor.[55]

Another factor benefitting Lysenko was that by 1936 his work in vernalization was being discussed seriously in Western scientific journals. Lysenko was described as a "leading investigator" who had developed "hypotheses of decided importance." Even if in some instances his interpretations were "too extreme," the fact that "laboratory experiments were very soon followed by extensive agricultural use" was taken as testimony to their value.[56]

---

[52] Kuznick, *Beyond the Laboratory*, p. 188.

[53] "Moscow Cancels Genetics Parley," *The New York Times*, December 14, 1936.

[54] "Quotations: Abandonment of the Moscow Meeting of the International Congress of Genetics," *Science* 84, no. 2190 (1936): 553–4; "Moscow Cancels Genetics Parley," *The New York Times*, December 14, 1936.

[55] Daniel Kevles, *In the Name of Eugenics: Genetics and the Uses of Human Heredity* (New York: Knopf, 1985), p. 118. "Nazi Sterilizations Had Their Roots in U.S. Eugenics," *Richmond Times Dispatch*, February 24, 1980.

[56] G.D. Fuller, "The Theoretical Significance of Vernalization; Vernalization and the Phasic Development of Plants," *Botanical Gazette* 97, no. 4 (1936): 867–8; Geo D. Fuller, "Vernalization," *Ecology* 17, no. 2 (1936): 298–9; D.J. Wort, "Vernalization of Marquis Wheat and Other Spring Cereals," *Botanical Gazette* 101, no. 2 (1939): 457–81; W.F. Loehwing, "Photoperiodic Aspects of Phasic Development," *Science* 90, no. 2346 (1939): 552–5; A.E. Murneek, "Recent Advances in Physiology of Reproduction of Plants," *Science* 86, no. 2220 (1937): 43–7; "Abstracts of Papers Presented Before the Physiological Section of the Botanical Society of America, Atlantic City, N.J. December 28, 1936 to January 2, 1937," *American Journal of Botany* 23, no. 10 (1936): 689; "Brief Notices," *The Quarterly Review of Biology* 12, no. 2 (1937): 232; During this same time period James G. Crowther (1899–1983), a popular science journalist in England, wrote that, "Vavilov considers the theory of the vegetative period is providing the most remarkable advances in plant physiology at

At a reception that same year Stalin proposed a toast saying that for science to flourish old leaders could not hide complacently in their shells. The future of science did not belong to the priests and monopolists—it belonged to the young—who did not want to be slaves and were bold enough to smash old traditions. An editorial in *Pravda* devoted his remarks to Lysenko.[57] In 1937 Lysenko was elected to the USSR Supreme Soviet. He appeared on the front page of *Pravda* as a political figure who proved that the lowliest peasant could become a great scientist. It was inevitable that now he would be seen standing up behind Stalin—applauding—performing as a leader.[58]

Lysenko's belief that nature would obey his will caused him to anthropomorphize as ardently as Michurin, and his initiatives were widely promoted. Lysenko believed self-pollination was degenerative and referred to cross-pollination as "marriage for love." As a result, hundreds of thousands of collective farm workers were sent into the fields to remove the anthers from spikes of wheat. Effectiveness was verified through questionnaires sent to the administrators which—whatever the actual results—came back positive.[59]

According to Lysenko, "survival of the fittest" was nothing more than the use of science to excuse economic exploitation. He countered: "The rabbit is eaten by the wolf but does not eat other rabbits; it eats grass. Likewise wheat does not crowd wheat out of existence." In reports of his work with *kok-sagyz*, a dandelion native to Kazakhstan valued for the high rubber content of its roots, Lysenko wrote how the plants cooperated with one another to defend themselves from weeds.

> In order that the weak *kok-sagyz* plants may be able not only to hold their own in this severe inter-specific struggle but to produce greater crops we have come to their assistance. … The weed attacks the hill but on encountering a mighty wall of resistance on the part of the numerous *kok-sagyz* plants it cannot make its way into the hill. And the *kok-sagyz*, having rid itself of its worst enemy, keeps on growing in bunches …

This method would receive its widest application in the cluster planting of trees during the Stalin Plan for the Transformation of Nature in 1949. When asked whether some weaker trees would need to be removed from the clusters to ensure their survival, Lysenko replied that no, these trees would sacrifice themselves for the good of the species.[60]

---

present and cites the work of T.D. Lyssenko [sic] as outstanding." Crowther also credited Lysenko with being the first to make vernalization a "general agricultural method." J.G. Crowther, *Soviet Science* (London: Kegan Paul, Trench, Trubner & Col, Ltd., 1936), pp. 287–8. In *The Lysenko Effect* Nils Rolls-Hansen also makes the point that Lysenko's theories fit in well with current research in plant breeding and physiology during this period. See particularly chapters 5 and 6.

[57] Joravsky, *The Lysenko Affair*, p. 105.

[58] Soyfer, *Lysenko and the Tragedy of Soviet Science*, pp. 99–100.

[59] Anna Bikont and Sławomir Zagórski, "Burzliwe dzieje gruszek na wierzbie," *Gazeta Wyborcza,* August 1–2 (1998): 12–5.

[60] Dominique LeCourt, *Proletarian Science: The Case of Lysenko*, trans. Ben Brewster (London: NLB, 1977), pp. 95–6.

## 2.4 Muller Flees the Soviet Union

Once Muller had left the Soviet Union for good Julian Huxley arranged a position for him at the Institute of Animal Genetics in Edinburgh, Scotland.[61] Muller confided grimly to Huxley about events in the land he had left behind. Muller did not want his words to be revealed publicly, and asked Huxley not to say anything to J.B.S. Haldane.[62]

> Haldane, especially, must not be informed—not now anyway—for I judge from the tone and content of his letters to me that he is at present having his political opinions impressed upon him with a rubber stamp (greatly as I admire his intellect and person) and would be influenced in the reverse direction from that which I intended.[63]

In the early 1930s, when Haldane was still at Cambridge, he had declared that though his department was full of communists, he was a Marxist—and the one he liked best was Groucho.[64] As the decade progressed Mussolini invaded Ethiopia, the Germans reoccupied the Rhineland and the Civil War erupted in Spain. Haldane traveled to advise the Spanish government on how to deal with gas attacks.[65] His wife, Charlotte Haldane, became a devoted Party member and traveled to China funded by the Communist International. J.B.S. remained uncommitted, however, and was primarily attracted to communism by the attitude that science should be useful rather than philosophical.

Haldane's scientific journalism was motivated by his belief that the public had a right to understand what went on inside the laboratories it paid to support. He published in the Sunday papers and popular magazines like the *Atlantic Monthly*, and explained that a writer should start from a known fact, such as a bomb explosion, bird's song or piece of cheese to illustrate a scientific principle.[66] This led him to the Marxist belief that the duty of a scientist was not just to explain the world but to transform it. By 1939 he was writing articles for the *Daily Worker* in which he explained that a photo of a demonstration by unemployed workers showed how human atoms act to change the structure of society.[67]

The Soviet Union was still, and would for a long time remain, an imaginary country to people like Haldane.[68] It was a place to project ideals and hopes that would prove hard to let go of. Muller was driven there by "radical" views, and

---

[61] Carlson, Genes, *Radiation and Society*, pp. 242–3.

[62] Diane Paul, "A War on Two Fronts: J.B.S. Haldane and the Response to Lysenkoism in Britain," *Journal of the History of Biology* 16, no. 1 (1983): 16.

[63] Adams, "Eugenics in Russia, 1900–1940," in *The Wellborn Science*, ed. Adams, p. 197.

[64] Clark, *JBS*, p. 107.

[65] Clark, *JBS*, pp. 138–9.

[66] Clark, *JBS*, p. 92.

[67] J.B.S. Haldane, "The Study of Change," *The Daily Worker*, June 29, 1939.

[68] Diane Paul makes this point in "A War on Two Fronts," p. 11.

now carried the troubling knowledge of experience. Muller was thus uniquely positioned to be hated by both sides—those who resented his disillusionment, and those who had thought he was wrong in the first place. As Muller wrote to Huxley, "in a world of conspirators, almost any action one can take becomes a conspiracy in one way or another!"[69]

In the Soviet Union Vavilov's position was becoming more precarious. In the spring of 1939 he told a meeting of his co-workers, "We will go the pyre, we will burn, but we will not renounce our convictions." On May 25 he presented a report on his institute's progress to the presidium of the Lenin All-Union Academy of Agricultural Sciences, chaired by Lysenko. Lysenko and his followers interrogated Vavilov on his ideas concerning the geographic origins of plant species. What did it matter if potatoes were brought to Russia under Peter the First? New varieties can be created in Moscow and Leningrad—that is the point, Vavilov was told. And why did he speak of Darwin rather than Marx and Engels? Vavilov tried to explain himself, saying he and Lysenko did not understand one another—though they discussed great things. Subtle games were being played, he said: "If Trofim Denisovich would only listen calmly instead of shuffling pages—life goes on…"[70]

Vavilov referred to the work of J.B.S Haldane to demonstrate that genetics could be Marxist: "Recently a book of Haldane's came out. He is an interesting figure, a member of the British communist party, an outstanding geneticist, biochemist, and philosopher. He wrote an interesting book entitled *Marxism and Science*, in which he tried…"[71]

Vavilov was interrupted by a member of the panel: "And got a dressing down."[72]

"Of course he got a dressing down in the bourgeois press, but he is so talented that he was admired even while being scolded…. He said that Marxism is more applicable to evolution, to history … that it can foresee much, just as Engels foresaw, 50 years ahead, many contemporary discoveries. I must say that I am a great lover of Marxist literature, not only of ours but of the foreign, too. There, too, many attempts at Marxist validation are made."[73]

Lysenko concluded the meeting by saying that Vavilov was "insubordinate": "I say that some kind of measures must be taken. We cannot go on in this way. … We shall have to depend on others, take another line, a line of administrative subordination."[74]

In June, 1939, a desperate group of geneticists addressed a letter to Andrei Zhdanov, head of the Ministry of Agitation and Propaganda, insisting that "conditions for work in the field of genetics are absolutely abnormal at the present time."

---

[69] Carlson, *Genes, Radiation, and Society*, p. 242.

[70] Medvedev, *Rise and Fall of T.D. Lysenko*, p. 62.

[71] Ibid, p. 63. Haldane did not actually join the British Communist Party until 1942.

[72] Ibid.

[73] Ibid.

[74] Ibid.

Lysenko and his followers were organizing a press campaign to abolish genetics once and for all. Devotion to Lysenko and Michurin had reached the point where people behaved as though nobody had ever discovered or been correct about anything before them. Zhdanov read their letter closely, and decided the matter should be judged in a public discussion at the Marx-Engels-Lenin Institute. The session, scheduled for the fall, would be run by *Under the Banner of Marxism*, a philosophical journal staffed by militant Marxists.[75]

The genetics congress to have been held in Moscow was rescheduled for Edinburgh, Scotland in late August, 1939. It was the eve of war. Vavilov was elected president of the congress but the Soviet delegation was refused permission to attend. Instead a letter arrived from Vavilov that he clearly had not written, but been forced to sign. Muller was the program chair and he read the papers of the Russian delegates in absentia, as other attendees stood around radios, hastily set up in meeting halls, to get the latest news.[76] The congress ended on August 31, the day before Nazi Germany invaded Poland. A dozen American delegates returned home on the *Athenia*, which became the first ocean liner to be torpedoed. Two geneticists from the University of Wisconsin went down with it.[77]

*Under the Banner of Marxism* held its meeting the first week in October. Vavilov understood that the purpose of the session was to "mutate" scientists into "Lysenkoites."[78] Another geneticist, Iulii Kerkis, challenged Lysenko to explain what he meant when he said that to obtain a result you must want to obtain it, and if you want to you will: "I need only such people as will obtain the results that I need."[79]

"I spoke correctly!" Lysenko responded.[80]

The geneticist said he could not understand how a scientist could just obtain what he needs: "It just doesn't fit in my head." [81]

Lysenko said that genetics was "too slow."[82]

Afterwards *Under the Banner of Marxism* judged that after years of tireless work Lysenko had produced practical results of great significance. He did not look upon science as "some precious rarity which exists only to be admired." As for Vavilov: "... our collective society has the right to make demands upon him..." Geneticists had proven themselves to be egregiously disdainful possessors of pontifical attitudes, cliquishness, aloofness, hostility to the new, and distaste for self-criticism. Disagreements and quarrels were necessary—scientists who refused them must be rebuked and excluded.

---

[75] Krementsov, *Stalinist Science*, pp. 66–70.

[76] Carlson, *Genes, Radiation, and Society*, pp. 264–5.

[77] Ibid.

[78] Joravsky, *The Lysenko Affair*, p. 109; The verb "mutate" comes from Vavilov. Joravsky coined the term "Lysenkoites."

[79] Joravsky, *Lysenko Affair*, p. 109.

[80] Joravsky, *Lysenko Affair*, p. 110.

[81] Joravsky, *Lysenko Affair*, pp. 109–10.

[82] Krementsov, *Stalinist Science*, p. 72.

Ten months later Vavilov was on a plant-collecting expedition in Western Ukraine—territory the Soviet Union had recently acquired from Poland thanks to the Hitler-Stalin Pact.[83] He and his research team drove three over-crowded cars into the foothills. The roads were difficult and, after several punctured tires, one of the cars was forced to turn back. On their return the members of Vavilov's group met a black Ford containing men in civilian clothes.

"Where did Vavilov's car go?" they asked, "We need him urgently." Vavilov's companions advised that the road ahead was treacherous, and said since Vavilov would be back by six or seven, it would be best to wait for him. "No," they responded, "we must find him right away, a telegram came from Moscow; he is being recalled immediately."[84]

Vavilov's car returned at dusk and was met by the men who had sought him earlier. Vavilov left with them, saying he would be back soon. Around midnight two of the men returned to say Vavilov would be flying back to Moscow. They showed a note Vavilov had written, instructing them to hand them over his things. The men were polite, if thorough, and did not initially object when two of the expedition members asked to accompanying them back to see Vavilov off at the airport.

Once outside at the car, however, they became sharp. There was not room in the backseat, and they asked—"Is it worth your while going?" Vavilov's companions assumed the men were joking, and one opened the rear door. He was punched and fell to the ground. One of the men ordered—"Let's go!"—and the car disappeared into the darkness.[85]

Vavilov was brought to Lubyanka prison and kept with 200 other prisoners in a cell meant to hold 25. One file in his dossier was labeled "Genetics," and contained three volumes of documents intended to prove he favored bourgeois science and opposed Academician Lysenko. After standing through repeated night-long interrogations his feet turned huge and gray. He began to crawl on all fours. At the trial he was sentenced to be shot, and transferred to Butyrki prison.[86]

As Nazi tanks advanced on Moscow Vavilov was moved again to Saratov prison. At times Vavilov brought a measure of order to things, as when he arranged a series of lectures on history, biology, and the timber industry among his fellow prisoners. But as time went on he became increasingly ill, and once was placed in a cell with a madman who stole his rations of bread. Vavilov was buried in a mass grave.[87]

---

[83] The agreement, officially titled *The Treaty of Non-aggression Between Germany and the Union of Soviet Socialist Republics*, divided Finland, Estonia, Latvia, Lithuania, Romania and Poland between Nazi Germany and the Soviet Union.

[84] Medvedev, *Rise and Fall of T.D. Lysenko*, pp. 72–6; The above account appears in Mark Popovsky, *The Vavilov Affair* (Hamden, CT: Archon Books, 1984), p. 127.

[85] Popovsky, *The Vavilov Affair*, pp. 128–9.

[86] Ibid, pp. 143, 155–6.

[87] Ibid, pp. 181–4.

## 2.5 World War II

H.J. Muller stayed in Edinburgh once World War II began. The previous May he had married again to Dorothea (Thea) Kantorowicz, a German Jew who was living in exile. Muller had left the Soviet Union and now his American citizenship made him a "neutral alien" in Scotland, unable to be out of doors after 8 p.m. This status also complicated his return home, requiring that he fly first to Lisbon, Portugal to catch a flight to the United States. He collected his most important papers and packed vials of flies four layers deep in a bread box. He and his wife went to London where they stayed with Julian Huxley. As they descended into an air raid shelter their first night, Huxley informed them one could always know the German Luftwaffe was approaching when the geese began honking—an instinct for enemies unchanged since Roman times.[88]

Once the Mullers left they found Portugal was a neutral zone where citizens and soldiers of countries at war wandered the streets together. It was 10 days before they could get a plane to the States and until then they lingered in the only hotel in town which would accept credit. It was also the most expensive. Muller needed to prepare food for the flies, but did not know the word for "yeast" in Portuguese. When Muller finally found some he left the bananas and yeast to ferment overnight, forgetting to take into account the warm climate. When he woke the next morning, Muller confronted a mess of reeking bananas, over-flowing vials and flies that needed transferring to new containers. He scrubbed the vials frantically in the common bathroom just off the lobby, shoving towels under the doors to contain the smell, afraid of being discovered and thrown out. The flies were the cause of drama again when they—his life's work—disappeared on the flight home. Muller pleaded with the steward to find them, and they were finally located, in the breadbox, where the bread was kept—in the pantry.[89]

For Dunn, the international tension leading to the Second World War spurred a new round of political activism. On February 13, 1939, he addressed students in New York as part of a city-wide program to combat racial hatred and ethnic bigotry, and in August he signed an open letter entitled, "To All Active Supporters of Democracy and Peace," to protest attempts to equate communism and fascism. In May, Dunn joined the National Emergency Conference to fight a number of bills before Congress which he argued could result in measures such as mandatory finger-printing, or concentration camps in the United States.[90]

In 1940 Columbia University president Nicholas Murray Butler announced that the Second World War was a conflict between "beasts and human beings," and any faculty member who disagreed should resign. Dunn led a protest and the demand was quietly withdrawn.[91] When American scientists acted to protect British research

[88] Carlson, *Genes, Radiation and Society*, p. 271.

[89] Carlson, *Genes, Radiation, and Society*, pp. 272–3.

[90] Kuznick, *Beyond the Laboratory*, p. 219.

[91] "Leslie C. Dunn, "Geneticist, Dies; Fought Racial-Difference Ideas," *New York Times*, March 20, 1974.

results that were threatened with war-time destruction, Dunn sent a group of white mice with genetic abnormalities he had received from J.B.S Haldane further west. He feared New York City was a location too at risk of imminent attack.[92]

Dunn also took on Father Coughlin, a Catholic priest whose anti-Semitic, anti-Bolshevik, pro-Hitler and pro-Mussolini radio broadcasts were wildly popular. On January 31, 1940, Dunn was both the chairman and featured speaker at an "Investigate Coughlin" meeting held at the Hotel Claridge in Times Square.[93] After the fall of France the *New York Times* published a letter from Dunn that described the "threat of execution now facing the brave men and women who were the first to oppose the Nazi and fascist tyrannies and who had thereafter to seek refuge in France."[94] He was also a founding member of the National Council of American-Soviet Friendship, created after a two-day congress in Madison Square Garden in November, 1942, which presented past Soviet achievements in science and medicine, and celebrated their military prowess.[95] Dunn and his wife also became joint treasurers of their local Russian War Relief fund, which that year collected nearly $2,000 in donations. Dunn was amazed at the public's generosity, particularly since the year before they had received almost no response aside from "caustic remarks."[96] In 1943, Dunn took on yet another responsibility by becoming president of the American-Soviet Science Society.[97] It seems likely that such a long list of activities would have left little time for scientific work.

Theodosius Dobzhansky, meanwhile, spent his summers conducting research in the San Jocinto Mountains of southern California. He released mutant flies into the wild and recaptured them with traps in the form of small cups full of fermenting mashed bananas on wire stands. Obtaining bananas was no easy feat, given war-time rationing, so sometimes fermenting dried banana flakes served as a substitute. Such activities, combined with Dobzhansky's thick foreign accent, soon attracted the attention of the authorities. He was questioned by the FBI, and the baffled agent ultimately decided to ask the advice of a nature counselor at a nearby Boy Scout camp who, after mulling it over, decided Dobzhansky was O.K.[98]

After Germany attacked the Soviet Union in June, 1941, Dobzhansky could no longer resist buying copies of the *Los Angeles Times* and *Los Angeles Examiner* from a nearby gasoline station. One of his colleagues favored Hitler and fascism, and since the Japanese had yet to attack Pearl Harbor, the war was still covered as a foreign conflict. Dobzhansky thought the papers were "dirty rags," and the "stupidity"

[92] "Reported From the Field of Science," *New York Times*, September 8, 1940.

[93] "Is Father Coughlin Too Big To Touch?" *New York Times*, January 29, 1940.

[94] "Letters to the Times," *New York Times*, July 27, 1940.

[95] Gormley, *Geneticist L.C. Dunn: Politics, Activism and Community*, pp. 243, 295–7.

[96] Ibid, p. 243.

[97] Ibid, p. 24.

[98] Coe in *The Evolution of Theodosius Dobzhansky*, ed. Adams, p. 23.

of the news combined with the opinions of those around him made his mood grim. Feeling the sentiment of an exile he envisioned the fall of Russia and establishment of a puppet government "under Hitler's boot." He wrote to Dunn that his imagination went even further, to the advent of a fascist government in the United States as well. It would be bad for everybody, but for people like them, "who committed the crime of being liberals," it would be worst of all. A foreigner and outsider he felt "powerless rage" and confided, "I almost prefer an outright Nazi with brains to these damn fools."[99]

As Hitler's army blockaded Leningrad the scientists at Vavilov's institute gathered un-ripened potato tubers from the experimental fields. They burned whatever they could to keep the collection from freezing in the cold darkness, and some starved to death rather than eat the seed-packets Vavilov had spent his life gathering.[100] Lysenko was awarded the Stalin Prize in 1941 and 1943, as the Soviet government's policies of internal secrecy and censorship kept biologists in the West from learning Vavilov's fate or being able to follow the situation in Soviet genetics. As Muller would later write, the war made it impossible to ever learn the actual cause of death of many distinguished scientists. All that seemed sure was they lived lives of terror—imprisoned, banished or executed.[101] Raissa Berg later described how when Moscow was under siege the residents rushed to grab incendiary bombs to toss them in sand before they could set buildings on fire. In contrast she recalled a colleague who was always the first to flee into any bomb shelter, followed by women nursing babies. That fact that this man later became an ardent follower of Lysenko's was, she believed, an "important little detail for understanding the fate of genetics" after the war.[102]

In 1941, 13 years after her first visit with J.B.S., Charlotte Haldane visited the Soviet Union again and asked an old friend if she could meet with Vavilov. "Vavilov? Vavilov? I do not remember what he is doing now. One has not heard of him for a very long time," she was told. When Charlotte returned to England she spoke to her husband about the changes she had observed since the last time they had been there. He refused to believe her and they soon divorced.[103] In 1942 Haldane finally joined the Communist Party and Vavilov was appointed a member of the Royal Society in London. The gesture was intended to not only to honor his accomplishments as a scientist, but also possibly to save his life. But it was too late.

---

[99] Correspondence, Th. Dobzhansky to L.C. Dunn, June 29, 1941. B: D917 L.C. Dunn Papers. Dobzhansky, Theodosius, 1940–1942. The American Philosophical Society.

[100] For a fictional account of this episode see Elise Blackwell, *Hunger* (New York: Little Brown and Co., 2003). See also, "Its Budget Slashed, Russian Seed Bank Fights for its Life," *New York Times*, March 23, 1993.

[101] H.J. Muller, "The Destruction of Science in the USSR," *Saturday Review of Literature*, December 4, 1948, p. 15.

[102] Berg, *Acquired Traits*, p. 70.

[103] Clark, *JBS*, p. 186; Haldane, *Truth Will Out*, p. 196.

J.B.S. Haldane's response to the war was as confusing and contradictory as his attitudes towards everything else. After the signing of the Hitler-Stalin pact he followed the "Party line" by arguing that the Soviets were forced to seek security since the allies, as capitalists, were unable to unite against Hitler. Haldane was also not pleased, as chairman of the editorial board of the *Daily Worker*, when the government shut down the publication's offices. However, he also began contributing to the British war effort.[104] Part of his work involved figuring out how sailors might escape from sunken submarines. Through this Haldane discovered that oxygen, at a pressure of five or six atmospheres, acquires the taste of ginger beer. It was he noted, "a trivial discovery which, for some reason, pleases me greatly."[105]

## 2.6 Between World War and Cold War

In New York Dunn conducted public lectures on "The Nature of Life" at the New School for Social Research and spoke at conferences of the American-Soviet Friendship Council.[106] He also took part in writing a pamphlet, "The Races of Mankind," which was to have been given to military officers by the War Department to educate American soldiers against the ideology of Aryan superiority. Distribution was blocked, however, by Chairman of the House Military Committee, Representative Andrew J. May of Kentucky, who objected to men from his state being informed that blacks were not, in fact, genetically inferior.[107]

Dunn's position at Columbia also brought him into proximity with advances that would soon heavily influence the relationship between scientists, government and society in the United States. The Manhattan Project occupied a portion of Schermerhorn Hall, just below Dunn's mouse laboratories. In the faculty lab he got a good deal of ribbing—"How many mutations did you get today? And how far are you from those atom-smashers down there? Does that have anything to do with it?"[108]

Once the end of the war drew near Dunn took part in public meetings to discuss postwar science policy. After the fall of Nazi Germany he spoke at a dinner sponsored by the National Council of American-Soviet Friendship at the Hotel Commodore. Generals George C. Marshall, Dwight D. Eisenhower and John J. Pershing, among others, sent messages praising the military achievements of the Red Army. Dunn read a paper on behalf of Harvard physiologist, Walter B. Cannon, arguing that shell-shock was much less common among the Soviets, because the

[104] Clark, *JBS*, pp. 146–7.

[105] Ibid, p. 162.

[106] "Is Father Coughlin Too Big To Touch?" *New York Times*, November 8, 1942; "Events Today," *New York Times*, November 6, 1943.

[107] "Letters to the Times," *New York Times*, March 14, 1944.

[108] The Reminiscences of L.C. Dunn, p. 994.

selfless "fighting spirit" of the Russian soldier was so strong that they did not care if they died.[109] This period when Dunn's beliefs and official attitudes towards the Soviet Union coincided would prove all-too-brief. In 1944 Muller declined Dunn's invitation to join the American-Soviet Science Society because he feared what effect, "the continuance of a pro-Soviet attitude ... so openly expressed," could have upon his career.[110]

On August 6 and 9, 1945, atomic bombs were dropped on Hiroshima and Nagasaki. Six days later the Japanese surrendered, and World War II was over. The Cold War would soon be underway. Scientists like Dunn and Muller, who had envied the support given to their colleagues in the Soviet Union, would now observe a new concern among scientists in the United States for the consequences and impact of their work upon government policy.[111] At the same time, Lysenko's dominance in Soviet biology would demonstrate the worst possible outcome when official interest became political interference in scientific research.

But where was Lysenko? When the war in Europe ended in May Red Square was filled with strangers who kissed, laughed and cried. Searchlights created a white ceiling for a sky lit by fireworks. In June the Russian Academy of Sciences celebrated its 220th Anniversary.[112] A number of distinguished foreign scientists were invited to attend the celebration, one of whom was Julian Huxley.

Huxley's delegation arrived first in Leningrad. They were given a small book, describing the history of the Academy, and two lists of members—dead or alive. Vavilov was on neither.[113] In addition to the expected tours of museums they visited a steelworks to witness the glory of Soviet production. Huxley noticed sentries posted outside the gates to the keep the workers, many of whom were political prisoners, from escaping. At a banquet the abundance of food was shocking, given the famine after the war. Their half-starved interpreters took the leftovers home to their hungry families.[114]

On the train to Moscow their delegation was greeted at each stop with official speeches and bouquets. They arrived to find a city overcrowded by destruction, with new examples of socialist realism replacing crumbling remains. They were taken to visit the Volga canal—a slave-labor project—and the house where Lenin died.[115]

The official opening session of the Academy's anniversary was held at the Bolshoi Theater. After a performance of Tchaikovsky's *1812 Overture*, the

---

[109] "Events Today," *New York Times*, December 15, 1944; "Red Army is Praised for Victories by U.S., British, Chinese Officers," *New York Times*, February 22, 1944.

[110] Gormley, *Geneticist L.C. Dunn: Politics, Activism and Community*, p. 341.

[111] See, for example, the activities of Manhattan Project scientists described in Wang, *American Science in an Age of Anxiety*, pp. 10–43.

[112] Berg, *Acquired Traits*, p. 110.

[113] The Reminiscences of Theodosius Dobzhansky, p. 174.

[114] Huxley, *Memories*, p. 269.

[115] Ibid.

*Marseillaise* and *God Save the Tsar*, the curtain parted to reveal a stage occupied by members of the Academy, surrounded by pink and white roses. As Huxley wrote, the price of the Great Patriotic War was evident in the gaunt faces of some, the spectacle of corpulent privilege obvious in others. When Huxley's delegation visited the Department of Darwinism at Moscow University, a red carpet was laid before them, while Soviet geneticists were told to use a separate entrance.[116]

Huxley's requests for an interview with Lysenko were turned down. He was suddenly informed however, that Lysenko would be giving a lecture the next morning at the Academy of Sciences. The hall was packed. Huxley attended with another British geneticist, Eric Ashby, and an interpreter. Ashby was the British scientific attaché in Moscow. Ashby had seen Lysenko previously and told Huxley he believed that Lysenko's tomato samples were not produced by cross-grafting, but rather had been carefully selected from a number of less valuable strains. This deception seemed even more likely given the fact that the tomatoes were only exhibited as wax models, and Lysenko did not bother to set up control groups or test the validity of his results with statistical methods.[117]

Raissa Berg, who was also in attendance, later recounted:

> Lysenko looked amazingly like Hitler. Even the lock of straight hair falling on his forehead was the same. The ability to exercise a hypnotic effect is one thing, attractiveness quite another.[118]

At one point during Lysenko's speech Huxley's translator was interrupted by a burst of laughter from the audience. Huxley was told Lysenko had mocked the theory of dominant and recessive traits and their segregation after crossing by claiming that dominance was the "digestion" of one "heredity" by another, while segregation of recessive traits in the second generation after crossing was "nature's belching."

> We know in our persons that digestion is not always complete. What happens then? We belch. So-called Mendelian segregation is nature's belching.[119]

On the last day of Huxley's visit a banquet was held, attended by Stalin. A table of hors d'oeuvres was laden with a boar's head and more vodka and caviar than Huxley had ever seen. Toasts went on for an hour and a half, and one was offered to Huxley, as the grandson of T.H. Huxley—"Darwin's great brother-in-arms." The program included a performance by a Red Army troupe—a band, singers, and tumblers turning somersaults, doing splits and dancing on bent knees like Cossacks. Stalin sat immobile beneath the stage, ignoring the performance. At one point Huxley tried to get closer to him but was escorted back to his seat by a waiter.

---

[116] Berg, *Acquired Traits*, p. 113.

[117] Huxley, *Memories*, p. 271.

[118] Berg, *Acquired Traits*, p. 114.

[119] Huxley, *Memories*, pp. 271–2; Berg, *Acquired Traits*, p. 114; Huxley, *Heredity East and West*, p. 102; In her memoirs Berg remembers the "nature's belching" comment as having been in response to a question by Huxley, however Huxley recounted it as being a part of Lysenko's address.

The waiter was courteous, but firm and—Huxley guessed—a member of the security police. The banquet did not end until after midnight and a few hours later he was half-asleep on a plane back to England.[120]

Dunn was also to have attended the celebration in Moscow as president of the American-Soviet Science Society, however at the last minute he had to cancel for medical reasons.[121] Several months later a State Department representative came to Columbia to ask questions about him. Dunn received a phone call from the president of the university and replied: "Well, why doesn't he talk to you first, and then I'll come over, or he can come and talk to me."[122]

The man arrived later in Dunn's office carrying a dossier from 1934 or 1935, and said—"We understand something about your record." He expressed surprise that a scientist like Dunn should be so concerned about what happened in other countries. It made sense that he would be interested in science and public relations, "That seems proper. But why does it have to be Russian? That's what puzzles us."

Strangely, during this same time period the State Department also asked Dunn if he would be interested in serving as scientific attaché at the American Embassy in London. Dunn said he would. The State Department then called and asked whether he had any objection to undergoing clearance. "Why no, that's up to you," Dunn replied, "How could I have any objection? That's something you're going to do. I'm not going to do anything about it."

The State Department asked whether he would be embarrassed if it should turn out he was not cleared. Dunn answered, "No, I'd like to find out the fact. That would be the most interesting outcome of the whole process. So go ahead."[123] It seems Dunn did not quite guess how quickly U.S.-Soviet relations were going to deteriorate over the next few years. His inability to grasp how his interest in, and enthusiasm for, Russian science would soon be used against him, was to have a disastrous impact upon his career.

***

This chapter began with a quote from a poem by German poet, playwright and director Bertholt Brecht, celebrating Lysenko. As you will see, Brecht was not the only artist who was interested in Lysenko's work. What about Lysenko's theories and career seems aesthetically appealing?

---

[120] Huxley, *Memories*, p. 274.

[121] The Reminiscences of L.C. Dunn, p. 752.

[122] In his oral history Dunn said that he received a phone call from the president of Columbia, however in 1945 Nicholas Murray Butler retired and for the next 3 years Frank D. Fackenthal served as acting president. The phone call Dunn received might have been from either Butler or Fackenthal. If it were from Butler his reaction would likely have been influenced by Dunn's protest against his demand, previously described, that faculty members who disagreed with his views on World War II should resign.

[123] The Reminiscences of L.C. Dunn, pp. 794–7.

Next you read about Muller's attempted suicide and emigration to the Soviet Union. Given that words like "crackpot" (attached to the term "pseudoscientist") would later be used (not infrequently) to describe Lysenko, is it not fair to ask if Muller's enemies and antagonists (real and otherwise) could not use this term to describe him as well? Is not trying to kill yourself typically regarded as evidence that someone has "cracked up?"

What about Muller's claim that capitalism is dysgenic? How is his argument similar or different from those made by Lysenko and "Lysenkoists?" Also, in what ways was the global economic crisis of the 1930s—the Great Depression—significant to this story? The fact that Marxism was being tested as an economic system even as capitalism seemed to have failed?

As for Vavilov's visit to Ithaca, review footnotes 9 and 15. Why do you think Dunn and Dobzhansky remembered certain details differently? How is this related to broader problems in historiography? Do you think either of them was being dishonest?

Was it wrong of Vavilov—given what he had told Dobzhansky—to invite Muller to join him in the Soviet Union? Why do you think he did? How about Muller's portrayals of Morgan and the "fly room" during these years? Is Muller's rhetoric any different than that of Lysenko et. al.?

Speaking of Lysenko, why did he choose the term "Michurinism" as a label for his theories? How is this relevant to the development of the term "Lysenkoism?" And what do you think of Vavilov's characterization of Lysenko ("angry species," "walked by faith, not by sight," "all progress in the world had been made by angry men")?

Is it not odd that at the same time Lysenko—who Muller regarded as a "pseudoscientist"—was on the rise, Muller's work was also being celebrated? How could it be that the climate of the USSR in the 1930s was such that both a future Nobel Prize winner in genetics, and one who would go down in the history of science as the man responsible for destroying genetics after WWII, could both be successful?

Thinking back to the slogan of collectivization quoted in Chap. 1 ("time does not wait"), how is Vavilov at a disadvantage vis-à-vis Lysenko during these years? And what about the letter Muller sends to Stalin along with the copy of *Out of the Night*? What was he saying?

Look back at the excerpt from Lysenko's address at the 1935 Congress of Collective Farm Shock Workers on p. 35. Who was he referring to? What was the strategic purpose of such statements?

What do you think Lysenko's relationship with his father was like? How many parents publish a letter on the front page of *Pravda*, thanking Stalin for making their son famous? But do you think they actually wrote it? Does it matter, or is it not the nature of the "selfish gene" to do whatever is best for its descendents? Also, how do ideas like "selfish gene" reiterate concepts that we see in the Lysenko affair?

Moving ahead to the relationship between Dobzhansky and Dunn, what do you think of their idea that American geneticists send a treatise to Moscow, written in "*their*" language, the "Marxist dialect," to defend genetics? How would the argument be phrased? How might this strategy backfire?

What about the news that Muller was in a "terrific emotional stew," and "willing to go to any foolish extreme to square himself" with Soviet authorities? Considering that Dobzhansky had made the decision never to return to the USSR, how would this sound to him? What did he think of Muller? Also, why did Muller not want Haldane to know what had happened?

Then there is the response of the Genetics Society of America. What do you think of Dunn's belief that the attack on genetics in the Soviet Union was analogous to the "Monkey Trial" in Tennessee? What were the similarities, what were the differences?

Dunn's involvement in politics also comes up in this chapter, and had an important impact upon his career as a scientist. One historian, Melinda Gormley, has gone so far as to coin the term "activist scientist" to describe Dunn.[124] Do you think this is the proper role for a scientist? What are the positives and negatives?

Also, looking back at Haldane's ideas about science and society, how was he similar to, or different from, Dunn? Do you agree that citizens have a "right" to understand the work carried out in laboratories they pay taxes to support? What about the idea that a scientist's role is not just to "explain" the world but to "transform" it? How does this relate to Dunn's scientific activism?

In this chapter I also described some of the content of Lysenko's theories, as well as how he and his followers portrayed the relationship between genetics and eugenics. To what extent was the criticism of eugenics legitimate? In what ways was it unfair?

As for Lysenko's ideas such as that self-pollination in plants is degenerative, intraspecific competition does not exist, and segregation of recessive traits in the second generation after crossing was "nature's belching?" What did he mean? And what about the criticism Vavilov received at the May 25, 1939 VASKhNIL session? Why were investigations of the genetic origins of plant species considered suspect?

As for the *Under the Banner of Marxism* session in October, 1939, shortly after the start of the Second World War, Lysenko was praised for not looking "upon science as some precious rarity which exists only to be admired," while Vavilov was told that in a state-funded science system "society" had the "right to make demands" upon scientists. What was meant by these statements, and how could one respond to them?

Walter B. Cannon, whose paper Dunn read at the dinner sponsored by the National Council of American-Soviet Friendship, had coined the term "fight or flight" three decades earlier.[125] How would this fit in with his claims about shell-shock and Russian soldiers? Does this type of analysis strike you as "scientific?"

---

[124] See Melinda Gormley, "Scientific Discrimination and the Activist Scientist: L.C. Dunn and the Professionalization of Genetics and Human Genetics in the United States," *Journal of the History of Biology* 42 (2009): 33–72.

[125] Walter B. Cannon, *Bodily Changes in Pain, Hunger, Fear and Rage* (New York: D. Appleton and Co., 1915).

That the relationship between science and politics in the U.S. was altered by the Manhattan project also comes up in the chapter. In what ways is this relevant to biology? How was the situation similar or difference for biologists versus physicists?

And then there was Huxley's visit to Moscow in 1945. Why do you think he pointed out, in his autobiography published several decades later, that Soviet geneticists were not allowed to walk the red carpet into the Darwin Museum, along with Huxley and his colleagues from the West?

# Chapter 3
# Reacting to Lysenko

*"Professor Muller's Flies"*

*Muller was a pupil of Morgan.*
*He laboured in the laboratory with tireless zeal. From the test tubes teeming with Drosophilae he expected an answer to the riddle of heredity, to the riddle of variability, the riddle of what controls forms, and many other riddles. ...*
*So Muller invented the queerest means of changing the hereditary nature of the winged captives in his test tubes. One day he put them under X-rays, and the flies which had been in the green spotlight of these rays brought forth unusual offspring. ...*
*Indeed, when turning his green spotlight upon his test tubes, Muller himself had no idea what would come of it. And when he obtained variations in his flies, he could not say why they changed in this way and not in another. It was like in the old fairy tale: "Go—I don't know where; bring—I don't know what."*
*And the idea began to creep into many minds that it may have been a mistake to repose these joyous hopes in the American flies that had been treated to X-ray shower baths.*

—from *Land in Bloom*, Stalin Prize 1949[1]

## 3.1 The Question of Heredity

In late October, 1946, the telephone rang while H.J. Muller was in the bathroom shaving. His wife answered and told him it was a reporter from New York who insisted on speaking with him. His face was covered in lather. "Damn it," he muttered, "what's the matter?" He grabbed the phone and barked angrily: "Muller!" He

[1] Safonov, *Land in Bloom*, pp. 126–9.

W. deJong-Lambert, *The Cold War Politics of Genetic Research: An Introduction to the Lysenko Affair*, Archimedes 24, DOI 10.1007/978-94-007-2840-0_3,

listened and replied quietly: "Are you sure? How do you know? Is it really true? Is there any confirmation of that? How can I believe that without any official notice?"[2]

He had won the Nobel Prize. Two months later in Stockholm City Hall Muller delivered his acceptance speech to King Gustav of Sweden and the other winners. He said they all "would be hypocrites if we were to pretend today that the increase of knowledge in any particular field inevitably leads to human betterment."[3] Muller also warned that the intellectual or economic enslavement of one group by another would be worse than the destruction of civilization itself.[4]

Muller was recognized for being the first to discover that X-rays produce mutations in *Drosophila*. A geneticist could treat flies with X-rays and then study their progeny to observe inheritance. Morgan and many other geneticists were initially skeptical, and Muller himself sometimes doubted what he had done. Mutations were described as "spontaneous," a term Dobzhansky referred to as a delicate way of saying you do not know what causes something. Lethal mutations were the most convenient kind to study because they eliminated the problem of some scientists being better at spotting mutations than others: Anybody can tell if something is dead or alive.

The practical value of Muller's work was that though X-rays could not be used to produce a specific type of mutation, they greatly increased their frequency. It was no longer necessary to sit around and wait for one to arise. The theoretical significance was that Muller had demonstrated that hereditary changes could be stimulated from the outside. With X-rays Muller partially fulfilled the promise of transforming nature. Humans could induce change, but not control it.[5]

X-ray mutations supplied possibilities that were both terrifying and hopeful. At a scientific conference in Washington D.C. in 1947, Muller told reporters that the descendants of Japanese people exposed to the atom bomb might suffer ill effects for centuries. If those that survived could see 1,000 years ahead, they might wish the bomb had killed them as well.[6] Muller also served on a commission to study the genetic impact the atomic blasts had on survivors' unborn children. Since the researchers understood the social stigma attached to giving birth to a malformed child in Japan was very high, and the population was suffering hunger from postwar shortages, they lured participants with food cards.[7]

Muller was not alone in his concern with the popular reception of scientific advances. He became part of a group of scientists headed by Albert Einstein who

[2] Carlson, *Genes, Radiation and Society*, p. 308.

[3] Carlson, *Genes, Radiation and Society*, p. 313.

[4] Ibid, pp. 313–4.

[5] The Reminiscences of Theodosius Dobzhansky, pp. 277–82.

[6] "Experts to Study A-Bomb Survivors," *New York Times*, September 1, 1947.

[7] Kevles, *In the Name of Eugenics*, p. 224; See also J.V. Neel and W.J. Schull, *The Effect of Exposure to the Atomic Bombs on Pregnancy Termination in Hiroshima and Nagasaki* (Washington, DC: National Academy of Sciences, 1956).

urged the formation of a world state. The only alternatives they could see were an expensive "armed peace," which would reduce living standards worldwide, or a devastating war between the United States and Russia. Scientists had a particular responsibility, they believed, because it was due to their work that civilization was now threatened.[8]

Scientists had become public authorities, but their authority, as the experience of Lysenko's critics would demonstrate, was not always trusted. Dobzhansky wrote to Dunn that he had heard Lysenko's situation was "less secure" than it had been. Soviet geneticists were hoping to "get out from under" him and requested the help of their American colleagues—Dunn in particular. Dobzhansky told Dunn he was considered "a sort of god" to them. He also had learned that Vavilov and several other geneticists "apparently do not exist any longer." "This letter is confidential," Dobzhansky wrote, "please do me the favor of destroying it."[9]

Like Muller, Dunn and Dobzhansky were becoming more aware of the necessity to educate the public on the implications of scientific advances. Their concern for how most people understood human evolution was expressed in a book they wrote together, *Heredity, Race and Society*, where they addressed the question of what, genetically speaking, the term "race" actually meant. Dunn first came up with the idea for the book after receiving a call from someone at WNYC, the New York City public broadcasting station. The station had been getting letters from members of the armed forces wondering why there was not anything on the radio about a topic that seemed so central to the war—racial difference. The Nazis were obviously using it one way, but was there another perspective? What were the facts? Dunn gave a series of broadcasts, and when he looked the scripts over later he realized that the idea of race, from the geneticists' point of view, had not really been openly discussed. He brought it to Dobzhansky and decided it was something they could do better together. Dobzhansky agreed.[10]

Dunn and Dobzhansky believed understanding why people are different was essential to world peace. Unfortunately the definition of race most people knew was anthropological, i.e., based upon visible difference. Geneticists defined races as populations of interbreeding groups. As Dunn put it, "the genetical definition is directed at the processes which produce races, and the anthropological definition is directed at the result." This was picked out by many readers as a new notion, and younger anthropologists seemed to welcome it.[11]

The book proved immensely popular. Dunn and Dobzhansky pointed out that since the Nazis believed being a Jew was hereditary they had sought to

[8] "Radioactive Rays Held Peril to Race," *New York Times*, April 2, 1947; "Atomic Scientists Urge World State," *New York Times*, April 12, 1948.

[9] Correspondence, Th. Dobzhansky to L.C. Dunn, July 4, 1945. B: D917 L.C. Dunn Papers. Dobzhansky, Theodosius, 1943–1945. The American Philosophical Society.

[10] The Reminiscences of L.C. Dunn, pp. 870–2.

[11] The Reminiscences of L.C. Dunn, p. 874.

exterminate them. Communism, on the other hand, they considered a matter of influence, and thus something which could be "cured" by indoctrination in a concentration camp.[12] Dunn and Dobzhansky criticized the hopes placed in eugenics, sterilization, and segregation. "Two thousand million similar robots, even happy robots," they said, "are not good material even for a world totalitarian state, much less for a society to live in."[13] Dunn and Dobzhansky also wrote that to claim that one group in human society is superior to any other is as silly as arguing wasps are superior to caterpillars because it is the former that stings and lays its eggs in the latter.[14] According to Dunn, each time the book was translated into a new language—Arabic, Hebrew, Chinese—Dobzhansky would say: "That's the best thing we ever did."[15]

A number of American geneticists, including Dunn, Dobzhansky and Muller, began translating and publishing the manuscripts of their Russian colleagues.[16] In England two biologists, P.S. Hudson and R.H. Richens, published a book, *The New Genetics in the Soviet Union*, which they intended as an impartial assessment of Lysenko's work. The tone of their analysis was unique in terms of the literature developing on the controversy at the time, and would not be heard again until over half a century later. Three themes ran through it: One, the difficulty of knowing what biologists in the Soviet Union *really* thought; Two, the difference between how Darwin's theories had been received in Russia versus the Europe and the U.S.; Three, the "alogical" stance adopted by both sides of the genetics debate in the USSR.

"Alogical" is a very interesting word. Here is their definition:

> The term "alogical" is used in this bulletin for methods of discourse other than logical. The term "illogical" usually implies a defective logical sequence, while alogical sequences are devoid of any sort of logical texture altogether. Conclusions reached by alogical discourse may be true or false, and if true are not demonstrably so without reference to subsidiary verification.[17]

Thus they placed "Lysenkoism" (a term which first appeared in the press the same year) at a place even lower in the hierarchy of knowledge than "illogical."[18] The previous two points—how the war and Stalinism had eclipsed contact between U.S. and Soviet scientists, and the enormous difference between how Darwin's reputation in Russia/the USSR had developed than elsewhere—are by now obvious in this story. Hudson and Richens was a valiant attempt, but their really serious error

---

[12] L.C. Dunn and Th. Dobzhansky, *Heredity, Race and Society* (New York: Pelican Books, 1946), p. 6.

[13] Ibid, pp. 11–2.

[14] Ibid, p. 114.

[15] The Reminiscences of L.C. Dunn, p. 880.

[16] Krementsov, *Stalinist Science*, p. 121. The other geneticists who participated were Isadore Michael Lerner, Ernest Babcock, G. Ledyard Stebbins, Walter Landauer and Jack Shultz.

[17] Hudson and Richens, *The New Genetics in the Soviet Union*, p. 23, ft. 1.

[18] See deJong-Lambert and Krementsov, "On Labels and Issues."

was believing that "underlying psychological motives" were what motivated behavior *only* in the USSR.[19]

Hudson and Richens identified four ways in which alogical discourse was enlisted by Lysenko and his followers: One, in appeal to a recognized authority (Dialectical Materialism, Darwin, Burbank, Michurin, Lysenko etc.);[20] Two, to claim that a given view is inconsistent with these authorities; Three, to imply something negative about the "state of mind of the author whose views are under consideration"; Four, to use practical utility as the sole criterion of truth.[21]

In the sections which followed—"Evidence," "Interpretation," and "Anti-Mendelism"—they provided an assessment of Lysenko's claims. Though, as mentioned above, the tone of their criticism differed sharply from the vitriol that appeared later, there was nothing in it that was positive. Hudson and Richens' summary of the evidence Lysenko produced for his theories was sprinkled with phrases such as—"practically useless," "not been proved," "little value," "not compelling," "lacks cogent demonstration," "no advantage" and "further experiments are needed."[22] Their review of the system upon which Lysenko based his ideas concluded that it was "lacking in cogency."[23]

What clearly bothered Hudson and Richens the most was Lysenko's attitude towards Mendel. Nevertheless, they used their account of his "anti-Mendelism" as an opportunity to call for a reconciliation between both sides of the debate.

> Lysenko's objection that other geneticists have refused to repeat his experiments remains, however, justified. He himself is, of course, open to the same charge, and his attitude has been so violently antagonistic to international genetics, that the resentful attitude of Mendelian geneticists is not surprising. It seems clear that little further progress can be made until a more accommodating attitude is adopted both by Lysenko and the international school of genetics. Some of Lysenko's results are certainly suggestive, but for the reasons already analysed they cannot be accepted without confirmation. If the industry displayed by members of both sides of the controversy could be expended in co-operative research, considerable progress might be made in such problems as heterosis and grafting phenomena. One of the principal objects of this bulletin has been to make possible such an approach to the subject. An attempt has been made to present all the evidence as it appears in the published writings of Lysenko and his school, and to analyse both the merits and the defects of his arguments. It is earnestly hoped that by so doing, much misunderstanding will have been removed and geneticists of each school will be encouraged to examine their own and each others (*sic*) data in an unprejudiced light. This should lead ultimately to a synthesis of what is best in both schools, thereby achieving that comprehensive understanding of genetical questions which is the aim of both bodies of investigators.[24]

---

[19] Hudson and Richen's cited Lenin's *Materialism and Empiriocriticism* as the source of this type of discourse.

[20] Another authority they mentioned was Kliment Timiriazev (1843–1920). To get a sense of the official line on Timiriazev's importance in Soviet biology see Georgiĭ Vasilevich Platonov, *Kliment Arkadyevich Timiryazev* (Moscow, Russia: Foreign Languages Publishing House, 1955).

[21] Hudson and Richens, p. 24.

[22] Ibid, p. 51.

[23] Ibid, p. 70.

[24] Ibid, p. 75.

The exact opposite of Hudson and Richens book appeared the same year: T.D. Lysenko's *Heredity and Its Variability*, as translated and published by L.C. Dunn and Theodosius Dobzhansky. In early spring, 1945, McGraw Hill sent Dunn a copy of the book, because they were considering it for publication.[25] Dunn brought it to Dobzhansky who agreed to do the translation.[26] The purpose of the translation was to introduce Lysenko to the American public and—as Dobzhansky put it—"let him stand on his own two feet."[27]

Dobzhansky would later say that Lysenko was foolish to not have hired a ghost writer. He was clearly not only illiterate scientifically, but literally as well: "His writings are undoubtedly actually his writings."[28] Dobzhansky referred to Lysenko's text as "excrement," and said the author himself was a "son-of-a-bitch": "Translating it has been one of the most unpleasant tasks I had in my whole life, and surely I would never undertake a thing like that for money—it can be done only for a 'cause'." If he could contribute even a little towards "unmasking this imposter," Dobzhansky said, it would be time well-spent.[29] As for Dunn, he believed that putting the book into circulation and having it judged on its merits just might turn out to be the most damaging thing they could do.[30]

Dobzhansky was also receiving letters from scientists who were refugees from the Soviet Union. They begged him to find work for them in the United States, no matter what it was. He confided:

> Oh, Dunn, what an indescribable tragedy has overtaken almost every one of my old colleagues and friends there! So many of them dead, and maybe after all this is the best for them; and now these people is this indescribable torture after the war has already ended, and when almost nobody will even look at them! You know, frequently I feel ashamed of my own welfare and relative happiness when I think of this mass suffering. And will there be time at long last when people will not escape from Russia like hunted animals and will not be slaughtered in Russia like cattle? It really seems that this country where I happened to have been born is holding the world's record of pitiless destruction of her own sons, and of her best sons at that. What an unenviable record.[31]

---

[25] T.D. Lysenko, *Heredity and Its Variability*, trans. Theodosius Dobzhansky (New York: King's Crown Press, 1946).

[26] Krementsov, *Stalinist Science*, p. 121. In his oral history memoir Dunn says that he spotted a copy of *Heredity and Its Variability* in a shipment of books sent to the American-Soviet Science Society (The Reminiscences of L.C. Dunn, pp. 747–8), however in *Stalinist Science* Nikolai Krementsov indicates it was sent to him by McGraw Hill.

[27] The Reminiscences of Theodosius Dobzhansky, p. 321.

[28] Ibid.

[29] Correspondence, Th. Dobzhansky to L.C. Dunn, July 31, 1945. B: D917 L.C. Dunn Papers. Dobzhansky, Theodosius, 1943–1945. The American Philosophical Society.

[30] The Reminiscences of L.C. Dunn, p. 748.

[31] Correspondence, Th. Dobzhansky to L.C. Dunn, August 20, 1945. B: D917 L.C. Dunn Papers. Dobzhansky, Theodosius, 1943–1945. The American Philosophical Society.

*Heredity and Its Variability* was published by King's Crown Press, a division of Columbia University Press, in 1946. In the preface Dobzhansky wrote: "The translator wishes to emphasize that his undertaking the work of translation does not imply agreement with the contents of the book, and that he reserves the right to criticize it has he sees fit." And this is exactly what Dunn and Dobzhansky did: They published Lysenko's book, and reviewed it themselves.[32]

## 3.2 Reviewing Heredity

Julian Huxley orchestrated reviews of *Heredity and Its Variability* in England, while Muller and Dunn wrote letters to colleagues that said negative reviews of Lysenko's work would weaken him. Reviews of *Heredity and Its Variability* appeared in all the major biology journals, including *American Naturalist*, *Physiological Zoology* and *Discovery*.[33] Dobzhansky's review was published in the *Journal of Heredity*.[34] Dobzhansky wrote that Lysenko's claims were, "to put it mildly, improbable," and that the same experiments conducted earlier by professional scientists had never produced the "startling results" Lysenko "alleged."[35] He also pointed out that in his opinion, (and contrary to the view of Hudson and Richens), these experiments were not worth repeating.

> Some people will probably wonder why geneticists do not rush to repeat these experiments. The answer is simple enough. The progress of science would be seriously disorganized if all scientists interrupt their work every time somebody publishes a dubious claim. Such claims are disposed of in due course. Admittedly, the history of science knows instances when claims first regarded as doubtful later proved to be valid and exerted an important influence on subsequent developments. One should not forget, however, that history is not as likely to record the vastly more numerous but less romantic instances when doubtful claims proved to be unfounded.[36]

Dobzhansky concluded by expressing confidence that Soviet geneticists would—despite Lysenko—continue to do good work. However, Dobzhansky also warned that it was "not Lysenko's fault that he has not succeeded in wrecking genetics as well as the agricultural sciences in the USSR," correctly guessing that, given proper support, he might be capable of doing so.[37]

---

[32] T.D. Lysenko, *Heredity and Its Variability*, trans. Theodosius Dobzhansky (New York: King's Crown Press, 1946).

[33] Krementsov, *Stalinist Science*, p. 122. See *The American Naturalist* 80, no. 790 (1946): 241–3; *Physiological Zoology* XV (1946): 332–4; *Discovery*, February, 1947, pp. 40–3.

[34] *The Journal of Heredity* 37, no. 1 (1946): 5–9.

[35] Ibid, p. 8.

[36] Ibid, p. 9.

[37] Ibid.

Dunn's review was published in *Science*, and the letter to the editor he wrote to accompany it gave no indication he had been involved in the translation. Dunn referred to "a book which has just been published" and received "very adverse criticism." Dunn justified his interest by mentioning that he was chairman of the American-Soviet Science Society, and had long felt it was important for the work of Russian scientists to be better known in the United States. Therefore his criticism could not be "attributed to animosity or prejudice" towards the USSR.[38]

The review itself gave the same impression of complete detachment, beginning,

> For a number of years biologists have been puzzled by the claims of the Russian agronomist, Lysenko, that Mendel's principles of heredity are all wrong and that acquired characters are inherited after all.[39]

Dunn noted the book was addressed to a general, rather than a "scientific and sophisticated" audience. Like Hudson and Richens, he referred to the lack of experimental data, and expressed hope that the results Lysenko claimed would be confirmed once they became available. Dunn also said that the book revealed a man whose purpose was different from scientists. Lysenko did not want to understand nature, but to control it, and condemned everything useless to this end. He was a "biological fundamentalist" whose scriptures read as if they were written in the nineteenth century. Dunn noted that Lysenko called Mendelian genetics "the pea law," and said he found Lysenko's "crude restatement" of outdated ideas surprising given the progress made in Soviet genetics from 1920 to 1940:

> It seems an anachronism somewhat like the denial of the facts of evolution over large areas of a country as progressive as the USA. In both cases the causes of such attitudes seem to those outside the country to be obscure and puzzling.[40]

Dunn was not content to just debunk Lysenko; he had other adversaries as well. Dunn was determined to use Lysenko to defend Darwinism from anti-evolutionists in the United States.

Shortly before Dunn's review was published, he received a letter from the Science Editor of the *New York Times,* Waldemar Kaempffert. "As you know," Kaempffert wrote, "Columbia University Press has published a monograph by Lysenko in which he expresses his well-known, low opinion of genetics as it was developed by Mendel and Morgan. I assume that this will create some stir." He wondered if Dunn could send him anything on the "Lysenko-Vavilov controversy," where "the pros and cons are objectively taken up."[41]

---

[38] Correspondence, L.C. Dunn to Editor, *Science*, January 1, 1946. B: D917 L.C. Dunn Papers. Lysenko Controversy in the U.S. #2. The American Philosophical Society.

[39] *Science* 103 (1946): 180.

[40] *Science* 103 (1946): 181.

[41] Correspondence, Waldemar Kaempffert to L.C. Dunn, January 29, 1947. B: D917 L.C. Dunn Papers. Lysenko Controversy in the U.S. #2. The American Philosophical Society.

Dunn referred Kaempffert to the reviews that he and Dobzhansky had written, and said he would probably want to consider the motives American geneticists had for publishing Lysenko. One point of view was that Lysenko's "vague and mystical ideas" did not deserve to be taken seriously:

> The other point of view rejects this as not conforming to the usual method of science which insists that what is criticized must be thoroughly understood first. Since I belong to this latter group, I believe that objective discussion of the scientific and practical bases of Lysenko's theories will eventually be worthy service to Soviet science.[42]

According to Dunn, the fact that the English translation of *Heredity and Its Variability* might be used as a "whip by those who wish to abuse the Soviet Union," was not as important as the long-term service which would be done to Russian science by exposing Lysenko.[43]

Kaempffert wrote back to Dunn gratefully: "Now that I have the opinions of my betters in genetics before me I ought to be able to handle Lysenko adequately." Kaempffert said he now had all he needed to explain the controversy to his "ungenetic" readers.[44]

In his review Kaempffert wrote that though it could be "jolted by X-rays" to create "two-headed calves and other monstrosities," the gene was a "Rock of Gibraltar." Genetics was "under fire" in the Soviet Union for pseudoscientific and ideological reasons. "Under the Marxian dispensation" heredity meant nothing—"if the environment is right, all is well with mice and men." Lysenko's theories were "pure Lamarckism," "something like believing that elephants grew trunks because they yearned for trunks." Lysenko was creating "too big a stir to be ignored," but Kaempffert believed genetics would survive the attack.

> All this is of some importance to American readers even though they have no great interest in what plant and animal breeders are doing with genetics in the way of producing faster race-horses, cattle which are immune to the tick, show-ring dogs that are all but abnormalities, sheep with longer wool. We are presented with an example of what State-imposed ideology can do to science. With Einstein's theory of relativity regarded as "bourgeois idealism" because the rejection of the infinite universe of Newton in some strange way violates the Marxian gospel, we have here another example of totalitarianism influence.

Unless Lysenko was read in this light, according to Kaempffert, his work just sounded like the ravings of a "crackpot."[45]

A few months later, in an article titled "Science and Ideology in Soviet Russia," Kaempffert made it clear to his readers why the Lysenko controversy mattered so

[42] Correspondence, L.C. Dunn to Waldemar Kaempffert, January 31, 1946. B: D917 L.C. Dunn Papers. Lysenko Controversy in the U.S. #2. The American Philosophical Society.

[43] Ibid.

[44] Correspondence, Waldemar Kaempffert to L.C. Dunn, February 1, 1946. Correspondence, Waldemar Kaempffert to L.C. Dunn, February 13, 1946. B: D917 L.C. Dunn Papers. Lysenko Controversy in the U.S. #2. The American Philosophical Society.

[45] Waldemar Kaempffert, "Man and His Milieu," *New York Times*, March 3, 1946.

much: the atomic bomb. Russia did not yet have the bomb, but they probably would soon. The Soviets claimed their success in science was due to Marxism, but Kaempffert claimed it was in spite of it. The need to think dialectically led to absurd studies such as "Marxism and Surgery," "The Dialectics of Graded Steel" and the "Dialectics of the Internal Combustion Engine." But there was more to it than zealotry, and no one could be sure where it was all going. Afterall, Kaempffert pointed out, despite the well-publicized purge in genetics, Professor Julian Huxley maintained a high opinion of Russian biology and many geneticists still practiced—for now.[46]

Dunn received a number of letters and telephone calls, particularly from college students, asking his opinion on Lysenko. Dunn thought one reason the controversy provoked so much debate was because of the appeal of Lamarckism, and the tendency of the general public to "grasp any straw that seems to confirm their almost innate desire to have evolution interpreted in this way."[47] Though Kaempffert tried to link Lysenko's appeal to Soviet socialism, Dunn believed that one did not have to be a Marxist to find the idea of perfectibility through adaptation to the environment attractive.

Dunn also received a letter from a colleague, Selig Hecht, praising his review of *Heredity and Its Variability* as "devastating," but said the final paragraph had left him "cold." Equating the fundamentalist minority in the U.S. with the situation in the USSR, Hecht argued, was unjust. Anyone who does not like the way biology is taught in their state can move elsewhere. Meanwhile for Vavilov the only place to go was a "concentration camp."

> Alas! The causes of such attitudes are <u>not</u> "puzzling and obscure." They represent people holding on to ancient history and ideas, <u>wishing</u> them to be true instead of facing new facts and ideas squarely and finding out how much truth there is in them.[48]

Still, Hecht wrote, he was glad Dunn was getting it all out in the open.[49]

Dunn himself did not necessarily agree with all the reviews of the book which appeared. He wrote a letter to the *Saturday Review of Literature* objecting to one reviewer's claims that not only did Lysenko represent the official Soviet doctrine in genetics, but that he was typical of Soviet science. "One should no more view the whole of Russian science through the lens of Lysenko than one should view American science through fundamentalist writings on evolution," Dunn responded. Meanwhile, Dunn himself was accused on the same pages of being a Soviet apologist whose agenda was suspect.[50]

---

[46] Waldemar Kaempffert, "Science—And Ideology—In Soviet Russia," *New York Times*, September 15, 1946.

[47] The Reminiscences of L.C. Dunn, p. 771.

[48] Correspondence, Selig Hecht to Leslie Clarence Dunn. Undated. B: D917 L.C. Dunn Papers. Lysenko Controversy in the U.S. #2. The American Philosophical Society.

[49] Ibid.

[50] Correspondence, L.C. Dunn to Editor, Saturday Review of Literature, undated. B: D917 L.C. Dunn Papers. Lysenko Controversy in the U.S. #2. The American Philosophical Society; "Letters to the Editor: Lysenko, Pro and Con," *Saturday Review of Literature,* March 30, 1946, p. 30.

Dunn had heard the Hearst Press empire, soon to play a central role in McCarthyism and the second Red Scare, was planning to use the book as anti-communist propaganda. Dunn wanted reviews that would claim Lysenko was an anomaly, but some of his colleagues objected to the idea of "organizing" reviews at all. L.J. Stadler at the University of Missouri was among those who refused Dunn's request. According to Stadler, a journalist interested in "axe-grinding" could just quote whatever part of the review he wanted. Aside from using Lysenko's book as a lesson on unscientific methods he did not see the value of translating him.[51]

## 3.3 Haldane and Heredity

Another solicited reviewer also declined, but for entirely different reasons. British geneticist J.B.S Haldane called Lysenko a "great Soviet scientist," and said that—all things considered—scientists received much more support in Russia than they did in the West.[52] When Muller heard that Haldane had refused he was outraged, however he also said that he had expected it.[53] Haldane's colleagues were well-aware of his personal evolution from socialist to communist, and they assumed this is what lay behind his refusal to criticize Lysenko.

Like Muller, Haldane preferred the unpopular side of any argument.[54] His support for Lysenko probably came from a desire to support the unorthodox, as well as the fact that he lacked Muller's recent first-hand experience with the situation in Soviet biology.[55] It was also, assuredly, the product of his resentment for Britain's reduced status as a global power, and dependency on the U.S., after World War II. Haldane's public statements at this point sound intended to provoke controversy, i.e., just what one might expect from Haldane. He excused the attack on Vavilov by saying "hard words break no bones," and compared his fate to an eminent British

[51] Correspondence, L.C. Dunn to Dr. L.J. Stadler at University of Missouri, December 22, 1945. Dr. L.J. Stadler to L.C. Dunn, December 28, 1945. B: D917 L.C. Dunn Papers. Lysenko Controversy in the U.S. #2. The American Philosophical Society; Ironically, 3 years later, the U.S. Department of Agriculture would decline to renew Stadler's passport to allow him to attend the Eighth International Congress of Genetics in Stockholm, due to suspicions concerning his loyalty to the United States. The Congress was held July 7–14, ending approximately one week before the VASKhNIL conference. [Finding Aid. Lewis John Stadler (1896–1954), papers, 1927–1955 (C2429). Western Historical Manuscript Collection-Columbia. University of Missouri/ State Historical Society of Missouri]. The State Department would decline to renew Dunn's passport, for similar reasons, in 1953. [Correspondence, Ruth Shipley to Leslie Clarence Dunn, April 9, 1953. B: D 917 L.C. Dunn Papers. Oral History Records. The American Philosophical Society]. See also below.

[52] J.B.S. Haldane, *Science and Everyday Life* (London: Lawrence & Wishart, Ltd., 1939), pp. 393–4; J.B.S. Haldane, "A Note of Genetics in the USSR," *Modern Quarterly* 1 (1938).

[53] Krementsov, *Stalinist Science*, p. 123.

[54] Clark, *JBS*, p. 128.

[55] See, for example, Haldane's *Callinicus*, cited above.

brain surgeon who had disappeared in the Middle East after protesting the treatment of skull fractures with brandy.[56]

Lysenko praised one of Haldane's works, and sent him a copy of *Heredity and Its Variability*. In January, 1945 Haldane wrote Lysenko a letter in response: "I should like to add that the book of mine to which you refer so kindly was written in 1941," Haldane said, "and that as a result, if I wrote it again today, I should modify my views in several respects, perhaps bringing them nearer your own."[57]

To some Haldane's views seemed more or less like Lysenko's already. A review of Haldane's book, *What is Life?*, in *The New York Times*, noted that though Haldane was peerless in his ability to present the complexities of modern science to the lay reader, the book was not recommended to anyone who did not follow the "party line."[58] Packed with gratuitous references to Marx, Engels and Lenin, the author wrote as though only a communist could think scientifically. The reviewer said Haldane viewed the Soviet Union as a "happy land" through "rose-tinted chiliastic glasses"; an ideal society where college professors can repair their own cars, and mechanics can discuss the scientific principles of their work?[59]

Haldane was far from the only prominent individual to support Lysenko. In a letter to Dunn, Dobzhansky described a meeting organized by the journal *Science and Society* where Irish playwright George Bernard Shaw praised Lysenko. Many of Shaw's plays dealt with philosophical questions related to malleability and the nature of human identity. *Man and Superman*—like Nazi belief in the Aryan "superman"—was inspired by Nietzsche.[60] *Pygmalion*—later made into the hit musical *My Fair Lady*—portrayed the gentleman Professor Higgins transforming the cockney-accented Eliza from a flower girl into a proper lady. Shaw's attitude distressed Dobzhansky because, as he wrote to Dunn, those who did not know better might assume "Lysenko is a great man, just because another great man says so!" However Dobzhansky also had a hard time taking Shaw seriously. The *Science and Society* affair was, he wrote, "indescribably funny," and declared—"Well life has some fun in it!"[61]

What disturbed Dobzhansky more was Dunn's sharpening radicalism. Timoféef-Ressovsky, who Muller had worked with in Berlin, had chosen to remain in Germany after Lysenko's rise to power during the 1930s. Timoféef-Ressovsky was captured in Berlin after World War II, a crime for which he was now, it seemed, going to be executed. Dunn believed Timoféef-Ressovsky deserved it for having betrayed the

---

[56] Clark, *JBS*, p. 197.

[57] Clark, *JBS*, pp. 199–201.

[58] J.B.S. Haldane, *What Is Life?* (New York: Boni and Gaer, 1947).

[59] Bernard Cohen, "J.B.S. and the Party Line," *New York Times*, December 28, 1947; "Chilias" is a term referring to the belief that Jesus Christ will someday reign on earth for 1,000 years.

[60] For an examination of Friedrich Nietzsche's influence on George Bernard Shaw see David Thatcher, *Nietzsche in England, 1880–1914: The Growth of a Reputation* (Toronto, ON: University of Toronto Press, 1970).

[61] Correspondence, Th. Dobzhansky to L.C. Dunn, January 10, 1946. B: D917 L.C. Dunn Papers. Dobzhansky, Theodosius, 1946–1947. The American Philosophical Society.

Soviet Union. Dobzhansky could not believe Dunn could be so cruel, especially since Dobzhansky himself had also chosen exile over Lysenko and Stalin. "Are you prepared," Dobzhansky asked, "to try and to electrocute all American citizens who happened to live in Germany during the war?"[62]

It pained Dobzhansky that Dunn was willing to give the benefit of the doubt to anything that happened in the Soviet Union. "Bloody despots of all times have maintained that they torture and kill for this or that very commendable purpose," he wrote. Dobzhansky also recounted a story he had heard of a Soviet official praising the efficiency of a concentration camp, saying that it "empties itself automatically into the grave." According to Dobzhansky, Dunn's criticism of Lysenko seemed too hesitant. He behaved at times as though there were shades of gray in an issue that Dobzhansky could only see as black and white.

> Are the "methods of discourse of Lysenko's school" justified by the fact that they live "in the midst of a society recently founded on Revolution?" Your statement that "some exponents of Mendelism were actually fascists" may be misinterpreted to mean Vavilov, although you surely do not mean anything of the kind. … Lysenko's power is surely granted him not by Russian peasants and workers. Peasants and workers at best found out about it post factum from the daily press.[63]

The tragedy was, Dobzhansky believed, that even if the horrors happening in the Soviet Union were true, the most vocal critics in the United States had unworthy motives for pointing them out. Dobzhansky saw around him propaganda which would soon be used to justify a war with Russia. The purpose of such a war would not be to liberate Russians, but rather to liquidate radicals, liberals and Reds in the United States and England by accusing them of high treason.[64] However the behavior of leftists like Dunn only made the situation worse.

> … to draw from all this the conclusion which you, evidently draw is also abhorrent to me. Namely you and men like you (whose general sympathies I certainly share) seem to feel that anything done in or by USSR must be if not actually defended then at least excused and given the benefit of every doubt (however far-fetched).[65]

In the end, however, Dobzhansky was willing to tolerate Dunn's attitudes out of friendship: "But I fear that we ought to have a real bull session," he said, "to try to understand and possibly to agree with each other on these philosophic-political matters."[66] Dobzhansky was also heartened by rumors that the situation in Soviet biology was altering in favor of genetics. A few weeks later he wrote to Dunn that he had recently heard that "Lysenko's star" was declining.[67]

---

[62] Correspondence, Th. Dobzhansky to L.C. Dunn, October 2, 1946. B: D917 L.C. Dunn Papers. Dobzhansky, Theodosius 1946–1947. The American Philosophical Society.

[63] Ibid.

[64] Ibid.

[65] Ibid.

[66] Ibid.

[67] Correspondence, L.C. Dunn to Th. Dobzhanky, November 25, 1946. B: D917 L.C. Dunn Papers. Dobzhansky, Theodosius 1946–1947. The American Philosophical Society.

On January 21, 1947, Haldane spoke at a Communist party rally in Madison Square Garden to celebrate the 23rd anniversary of Lenin's death. Dobzhansky regaled dinner guests in his apartment with the story of the night he had arrived in Petrograd and Lenin died: "Well surely the devil is taking the soul of some awful sinner to hell." A guest jumped up on her chair and scolded him. It is a shame Haldane did not hear the story, Dobzhansky wrote to Dunn, he could have told it to Madison Square Garden.[68]

## 3.4 VASKhNIL

In the summer of 1945 Soviet geneticist, Anton Zhebrak, came to the United States as part of a UN delegation. Dunn knew Zhebrak, having helped arrange his appointment to study at Columbia in 1929, two years after Dobzhansky's arrival. At the time Zhebrak had been fascinated by theoretical genetics, but Dunn decided to put him to work on experimental jobs after determining he could use the experience. Zhebrak, however, did not seem interested. Dunn had the impression that Zhebrak must have been let out of the Soviet Union because he was trusted by the Party, not because he was thought of highly by geneticists. After Dobzhansky's defection the Soviet government had become more cautious about the loyalty of the scientists they sent overseas with the Rockefeller Foundation. In the end Dunn doubted that Zhebrak had "acquired very much additional knowledge" during the time he spent in his laboratory.[69]

Regardless of Dunn's assessment, Zhebrak had since risen to become one of the most influential figures in Soviet genetics. He was appointed president of the Belorussian Academy of Sciences where he organized a new genetics laboratory.[70] On his return visit to the United States after the war Zhebrak attempted to meet with Dunn and Muller, however due to last-minute changes he was forced to return to Moscow immediately via Alaska. During his short stay, however, Zhebrak was confident enough to predict to Western geneticists that it would not be long before Lysenko had enough rope to "hang himself."[71]

In the wake of Zhebrak's visit, a group decided to once again try and organize an international genetics congress in the Soviet Union. As a first step they chose

[68] Correspondence, Th. Dobzhansky to L.C. Dunn, January 23, 1947. B: D917 L.C. Dunn Papers. Dobzhansky, Theodosius 1946–1947. The American Philosophical Society.

[69] The Reminiscences of L.C. Dunn, pp. 754–6.

[70] Krementsov, *Stalinist Science*, p. 146.

[71] Correspondence, Th. Dobzhansky to L.C. Dunn, July 4, 1945. B: D917 L.C. Dunn Papers. Dobzhansky, Theodosius, 1943–1945. The American Philosophical Society; Correspondence, I.M. Lerner to H.J. Muller, June 25, 1945. Correspondence, I.M. Lerner to L.C. Dunn, June 27, 1945. Correspondence, I.M. Lerner to H.J. Muller, June 29, 1945. B: L563 Isadore Michael Lerner Papers. Newspaper – Z. The American Philosophical Society.

Zhebrak as Soviet representative on the organizing committee to replace Vavilov.[72] In October Zhebrak published a piece in *Science* challenging criticism of the current state of Soviet biology—Lysenko in particular. He argued that while Lysenko had been rewarded for his practical contributions in agriculture, many geneticists had also been decorated for their work by the Soviet government: "These facts should serve to show that Academician Lysenko's criticism of genetics, based as it is on naïve and purely speculative conclusions, despite the vigor of its assault is incapable of impeding the onward March of genetics in the USSR."[73]

Plans to hold a congress in Moscow ultimately failed, but Stockholm was chosen as an alternative because of Sweden's proximity to the Soviet Union. However, in the fall of 1947, Zhebrak was criticized in *Pravda* for his article in *Science*. He was charged with humiliating and defaming Lysenko, while failing to mention other great figures in Soviet science, such as Michurin. Zhebrak was forced to defend himself in a court of honor and relinquish his position with the Belorussian Academy.[74]

Meanwhile, the Soviet Union was weak and starving. Nearly 26 million citizens had died during the Second World War, and now faced possibly the worst famine in decades. Stalin summoned Lysenko to the Kremlin on New Year's Eve. They discussed wheat production, and Lysenko subsequently received 200 kg of branched wheat seeds to study. Lysenko kept Stalin regularly informed of the results of his work with the seeds. He also took the opportunity to complain of the harm done to agriculture by genetics. Geneticists, he said, slandered Michurinism. Lysenko urged Stalin to intervene, but Stalin waited.[75]

In the spring of 1948 a member of the politburo, Iurii Zhdanov, gave a lecture entitled, "On Issues of Modern Darwinism." Iurii's father Andrei, organizer of the last debate between Vavilov and Lysenko, had been an important figure in the purges of the 1930s. Lysenko did not attend Iurii Zhdanov's lecture, but listened to it on the public address system, sitting in the office of an ally, Mark Mitin.[76] Zhdanov criticized Lysenko by insisting living organisms of the same species did indeed seek to eliminate one another: Intraspecific competition was a fact. Zhdanov also attacked Creative Darwinism, and would soon accuse Lysenko of sabotage.[77] Lysenko, meanwhile, responded in kind. When he learned that Ministry of Agriculture planned to introduce a tetraploid variety of *kok-sagyz*—the precise organism he used to illustrate

---

[72] Krementsov, *Stalinist Science*, p. 123.

[73] Zirkle, *Death of a Science in Russia*, p. 59. See also Anton Zhebrak, "Soviet Biology," *Science* 102, no. 2649 (1945): 357–8.

[74] Krementsov, *International Science Between the Wars*, pp. 141–2; Krementsov, *Stalinist Science*, pp. 120–4, 146–8.

[75] Krementsov, *Stalinist Science*, p. 160.

[76] Ibid, p. 164.

[77] The use of the term "Creative Darwinism" by Lysenko and his followers was a way of claiming priority over evolutionary theory at a point when genetics was being established as the science of evolution. See Krementsov, *Stalinist Science*, pp. 149–55.

his belief that intraspecific competition was a myth—Lysenko termed it a "genetic monster," and charged that the "Mendelist-Morganists" had "literally thrown aside all restraints."[78]

At the international genetics congress in Stockholm in July Muller was president, and not a single Soviet geneticist was in attendance. Later that month Stalin requested Lysenko write a report on the conflict with genetics and send it to him. Stalin edited the text and Lysenko presented his report on the last day of July at a session of the Lenin All-Union Academy of Agricultural Sciences, at the Ministry of Agriculture in Moscow.[79] The session lasted a week, and for the first part only Lysenko's supporters were given the floor. There were nearly 700 attendants—Michurinists, geneticists, philosophers and powerful bureaucrats. Lysenko's report was titled, "On the Situation in Biological Science," and the next day, Sunday, the conference participants went to Lysenko's model farm in the Lenin Hills nearby.

Lysenko described two worlds, two ideologies in biology, and informed his audience they were irreconcilable.[80] The difference between Weissmanism-Mendelism-Morganism and Michurinism had now ignited, he said, over the question of whether the conditions of life influence the living body. Lysenko mocked the fly researchers, citing the work of a geneticist, N.P. Dubinin, who had concluded that the chromosome structure of the fruit fly population of the city of Voronezh had shifted as a result of the Second World War: "Such destruction is called selection…," Lysenko said, and the audience laughed. According to Lysenko, Dubinin was more concerned with flies than with the people who had sacrificed their lives for their country: Such is the Morganist contribution to science.[81]

The alternative, Lysenko said, was Michurinism. Darwin's theory was insufficient on its own. Michurin added the understanding that man can alter evolution in the direction he desires. Lenin and Stalin "discovered" Michurin and "made his teaching the possession of the Soviet people."[82]

> Our Soviet, Michurinist Darwinism is a creative Darwinism, raising and solving the problems of the theory of evolution in a new way, in the light of Michurin's doctrine.[83]
>
> Until now, however, "Morganism-Mendelism" had dominated the curriculum in Soviet colleges in universities, while the revolutionary doctrine of Michurinism "fostered by the Bolshevik Party and by Soviet reality, remains in the shade."[84]

---

[78] Krementsov, *Stalinist Science*, p. 164.

[79] For a description of Stalin's edits see Krementsov, *Stalinist Science*, pp. 159–69, Kirill O. Rossianov, "Editing Nature: Joseph Stalin and the 'New' Soviet Biology," *Isis* 84, no. 4 (1993): 728–45, Pollock, *Stalin and the Soviet Science Wars*, chapter 3.

[80] The quotations which follow come from the official transcript, published in English by the Foreign Languages Publishing House in Moscow. An alternative translation, carried out by colleagues of Conway Zirkle, is described below.

[81] *The Situation in Biological Science: Proceedings of the Lenin Academy of Agricultural Sciences of the U.S.S.R., Session: July 31–August 7, 1948, Verbatim Report* (Moscow, Russia: Foreign Languages Publishing House, 1949), pp. 31–3.

[82] Ibid, p. 49.

[83] Ibid, p. 47.

[84] Ibid, p. 44.

Over the next week Lysenko's followers presented one by one. Genetics, they said, was an anti-people trend whose purpose was to transform Soviet citizens into passive spectators, waiting indifferently and submissively on the kindness of nature.[85] Geneticists were singled out for attack, accused of attempting to separate science from the Soviet people. Party philosopher, Mark Mitin, commented:

> For many years the representatives of the Mendel-Morgan trend have been engaged in fruitless laboratory experiments divorced from life, divorced from the requirements of the people and of socialist construction. It is an anti-popular trend in science.
>
> To what disgusting monstrosities this trend leads, was shown in his address by T.D. Lysenko, when he quoted as an example the researches of Dubinin on the influence of our Great Patriotic War upon the chromosome apparatus of fruit flies.
>
> The name Dubinin deserves to become the synonym for divorcement of science from life...[86]

Theodosius Dvoryankin, another follower of Lysenko's, said the idea of pure hereditary lines was "reactionary and preposterous piffle," intended to rationalize European monarchy.[87] Lenin instructed that the deductions of bourgeois professors who see nature through the eyes of elitism must be ignored. Genetic notions such as selection, Dvoryankin said, are simply reflections of capitalist ideology projected onto nature.[88] The existence of genes was once again declared to be "utter fiction."[89]

> Geneticists attempts to alter these fictitious genes were portrayed as crude approximations of Michurinism: "...variability may be of different kinds: you can kill an organism with a stick, the organism will suffer a change, but there will be no development..."[90]

Lysenko's followers responded to the accusation that he was a Lamarckist, by praising Lamarck. They said geneticists considered Lamarckism a regressive, outdated interpretation of evolution, because it threatened the bourgeois capitalists system.

> As is known, Lamarck's theory arose in connection with the ideas of the French encyclopaedists and the French materialists. It reflected the revolutionary epoch of that time. ... The reaction against the French Revolution also caused a strong reaction against the ideas of Lamarck...[91]

---

[85] Ibid, p. 253.

[86] Ibid, pp. 263–4.

[87] Details of the official English translation of the VASKhNIL session will be covered below in the discussion on Conway Zirkle's *Death of a Science in Russia.*

[88] Ibid, pp. 362–3.

[89] Ibid, p. 325.

[90] Ibid.

[91] Ibid, p. 273; For an analysis of the relative importance of Lamarckism versus Darwinism in the Lysenko controversy see Alexei Kouprianov, "The 'Soviet Creative Darwinism' (1930s–1950s): From the Selective Reading of Darwin's Works to the Transmutation of Species," *Studies in the History of Biology* 3, no. 2 (2011): 8–31.

Geneticists were thus positioned as regressive absolutists who feared "enlightenment."

In addition to the notion that Lamarck's theories were the product of revolution, Lysenko's followers provided further evidence for the belief that acquired characteristics are inherited. They argued that data from collective farms demonstrated that the exercise cows' udders received from milk maids during milking significantly increased output:

> It has been established that in order to obtain one litre of milk the milkmaid must produce over a (*sic*) 100 squeezes with her hand. A cow yielding 6,000 kg. of milk would therefore be subject to more than 6–7 million such irritations during her lifetime. …
>
> The cow's udder, which is one of the most important parts of her organism, changes gradually under the influence of our action, which in turn causes corresponding changes in the entire milk forming apparatus. Step by step the cow's organism is changed and adjusted to the requirements made by man upon the cow's udder with indefatigable and ever increasing insistence. …
>
> We state on the basis of many years of observation that all these changes in the udder of a milch cow are a direct result of our external actions. Photographs of the udders of Amazonka and Barkhotka show very distinctly the character and strength of the influences we exert.[92]

Just as Soviet cows were enlisted in support of Lysenko, the machines used to milk cows in the United States were cited to highlight the corrupt practices of capitalism. The way they were designed supposedly ignored the physiological needs of the cow.

> Capitalist technique has brought forth milking machines. They are two-beat machines. The machine produces the beat of the squeezing of the cow's teat and the beat of the sucking of the milk. … If a milking machine is used on a cow suffering from mastitis in a latent form the ailment will become more severe.[93]

The same was true of plow production. Though the Soviets had settled upon a design that was highly regarded worldwide, various firms in the United States competed to produce a needless plethora of plow types. The result was fields choked with weeds and exhausted soil.[94]

Though inferior agricultural technology in the U.S. was an important plank in the Lysenkoist platform, the practice and testimony of peasants and agricultural workers was the most important evidence on behalf of Michurinism. Lysenko's followers said geneticists had no idea what these heroes of labor had achieved. They too had opinions on topics that geneticists considered "academic controversy," and knowledge was power: "No one can now succeed in robbing them of this new outlook on nature, no one can strike from their hands the weapon which makes them transformers of nature."[95]

---

[92] Ibid, p. 257.

[93] Ibid, pp. 534–5.

[94] Ibid, pp. 530–1.

[95] Ibid, pp. 433–423.

At the VASKhNIL session genetics was presented as a stale and reactionary doctrine, the exact opposite of the vibrant science of Michurinism. When geneticists were able to respond—still under the impression they were there to discuss and debate—they were at a loss. All the words they would have used to defend themselves had been turned into terms of criticism; the grounds upon which they would have justified their work had been yanked from beneath them.[96] Zhebrak rose to defend genetics by citing practical results, but Lysenko interrupted him, saying it was work he and his followers had already achieved: "What you are telling us is quite a common matter," he said.[97]

Later one of Lysenko's followers, Vsevolod Stoletov, also mocked Zhebrak's presentation.

> ...Professor A.R. Zhebrak yesterday exhibited from this platform sheaves of wheat which we have been seeing for about three years. ...
>
> These sheaves were from the 1945 harvest. Is it that the experimenter has grown nothing new in this period?[98]

Stoletov went on to state that Michurinists had also demonstrated the inheritance of acquired traits in *Drosophila*, and proven that hybridization was not a mere "shuffling of permanent genes," but a method for creating "new forms of plants ... possessing properties which neither of the parents possessed."[99]

Other geneticists attempted to offer a compromise or locate an escape. B.M. Zavadovsky said the organization of the session had "not been quite normal," and he had only received a last-minute invitation while in Moscow on his way from one sanatorium to another. Zavadovsky said that while he agreed with the whole "line of attack" against "formal genetics," he did not believe it was fair for all geneticists to be lumped in with the "Weismannist-Mendelists."[100] He did not believe either that there were only "two trends" in Soviet biology, or that Michurinism could account for everything.[101]

Another geneticist, S.I. Alikhanian, also attempted to disassociate genetics from the "idealistic theories of Weismann."[102] He updated the audience on more recent work in genetics by quoting a paper of Muller's, ("I hope I shall not be accused of fawning if I permit myself to cite the American geneticist Muller"), which referred to the role of environmental conditions in the development of the organism.[103]

---

[96] Nikolai Krementsov points out that the order of speeches was carefully arranged so that only Lysenko's allies were given the floor during the first half of the meeting. Krementsov, *Stalinist Science*, p. 169.

[97] Ibid, p. 472.

[98] Ibid, p. 566.

[99] Ibid, pp. 567–8.

[100] Ibid, pp. 334–5.

[101] Ibid, pp. 338, 347.

[102] Ibid, p. 427.

[103] Ibid, p. 429.

Alikhanian concluded by proclaiming that Soviet biologists must unite against criticism from the West, in the spirit of their shared belief in the "one and only progressive doctrine—the teaching of Lenin-Stalin."[104]

## 3.5 Finale

The session went on like this for days. Speeches were interrupted with prolonged applause. Praise for Lysenko alternated with continual references to the glory of Soviet science, until the two became indistinguishable. Meanwhile, genetics was tainted by the invective hurled at Mendelist-Weissmanist-Morganists. On the second-to-last day, August 6th, the philosopher I.I. Prezent, Lysenko's closest ally, spoke.[105] Prezent asked, rhetorically, whether there was any possibility for reconciliation between geneticists and Michurinists. He answered his own question by joking that the former were only satisfied by statistics: the audience laughed. Prezent then made it clear that the only choice for geneticists was to surrender their belief in the "mythical," "false idea" of a "hereditary substance," i.e., the gene.[106] To believe in the gene, Prezent said, "one must really be a Dubinin"—and the audience applauded: The term "*dubina*," in Russian, meant "blockhead."[107]

Prezent then divided the genetics camp between those who refused to compromise, and those who attempted to reconcile genetics and Michurinism. Who, he asked, should be considered "most dangerous to the progress of the Michurinian theory," those who like Dubinin, Zhebrak and "their ilk" defended genetics, or "people of the type of Zavadovsky and Alikhanian" who "smuggle in anti-Michurinian, Weismann-Morganian views under the guise of sympathy for Michurin."[108] According to Prezent, the latter were far worse. He compared Zavadovsky's equivocations with a museum display he had observed before the war. One side of the exhibition boards showed the Michurin view of evolution, and the reverse side displayed Morgan's: "...depending on the composition of the visiting group, it was possible to turn these exhibits any way you choose."[109]

---

[104] Ibid, p. 440.

[105] The partnership between Lysenko and Prezent dated back to 1934. Prezent was the one primarily responsible for giving Lysenko the language he needed to present his theories as Marxist science. Krementsov, *Stalinist Science*, p. 59.

[106] Ibid, pp. 576–7.

[107] Ibid, p. 577; Zirkle, *Death of a Science in Russia*, p. 239.

[108] *The Situation in Biological Science*, p. 577; Nikolai Krementsov has pointed out that most of those criticized by name at the conference were not geneticists. Of these four only Zavadovsky was not a specialist in genetics, he had however, been an active critic of Lysenko (Krementsov, *Stalinist Science*, p. 171).

[109] Zirkle, *Death of a Science in Russia*, pp. 244–5; This story was left out of the official transcript. See below.

Reconciliation was now hypocrisy and deception. Prezent portrayed the position of geneticists as desperate: "… they are thinking up, in addition to the gene, all sorts of 'plasmogenes,' 'plastidogenes' and similar terms designed to veil the complete theoretical and factual discomfiture of Morganism."[110] He complained that in Hudson and Richens' review of Lysenko's work, the authors had written Lysenko's results were "unsound" because, "as they claim, everyone knows that the Russian varieties are impure." Hudson, Richens and other "Morganists," "must be in a pretty bad fix," Prezent said, to be making such arguments.[111]

> Different beings resort to different means of defense: the lion defends itself with its claws and the bull with its horns, the hare relies upon the swiftness of its legs, the mouse hides in a hole, and the cuttlefish—it secretes dark fluid and escapes from its enemy into the murk. Our anti-Darwinists are fond of imitating the tactics of the cuttlefish, with the only difference that the latter is of course glad if it can only get away from its enemy, but our Morganists hurl abuse at their opponent from out of their murky cloud, and complacently cry: "We've smashed him! We've won! He's crushed!"[112]

Prezent characterized genetics as "alien," "brought to our country from foreign shores," and said it required the same blind faith as religion.[113] He accused geneticists of making false analogies between the "invisible atom" and the "invisible gene": "Far closer would be an analogy between the invisible gene and invisible spirit."[114] Prezent claimed the "Morganists" wanted to discuss their differences with Michurinists, but "we shall not discuss … we shall continue to expose them."[115]

Prezent finished by accusing geneticists of portraying outdated aspects of Michurinism as contemporary theory. This was, in fact, the very tactic Lysenkoists frequently used against geneticists: Weisman's belief in strict inheritance was attached to current genetic research by calling geneticists "Weismanists" who practiced "Weismanism," while the role of negative eugenics in Nazi ideology provided the label "fascist science." Along these same lines Prezent concluded:

> The Morganists are endeavoring to hold back the Michurinian doctrine by contrasting Michurin to Lysenko, the earlier Lysenko to the later Lysenko, and Lysenko to his followers. That is what retrogrades may be expected to do. They know that every new advance spells their ruin.[116]

Prezent's presentation unwittingly summarized the content of what the term "Lysenkoism" was encompassing: Sarcasm as scientific discussion, intolerance as dialogue. The remaining ingredient was the exercise of power.

---

[110] *The Situation in Biological Science*, p. 596.

[111] Ibid, p. 584.

[112] Ibid, p. 600.

[113] Ibid, p. 603.

[114] Ibid, p. 602.

[115] Ibid, p. 603.

[116] Ibid.

Prezent gave his speech on August 6, the second-to-last day of the conference. According to an account Prezent gave later, Lysenko met with Stalin that evening.[117] Stalin asked Lysenko: "How would you tell people that the Central Committee of the Party approved your report?"

"I do not know, I could not say," Lysenko replied.

"You can. Take a pencil and write."[118]

Stalin dictated the opening paragraph of Lysenko's speech to make it sound as though some anonymous individual asked him what the position of the Party was on genetics. It was choreographed authenticity: A simple question, an honest answer. Lysenko delivered it the next morning:

> Comrades, before I pass to my concluding remarks I consider it my duty to make the following statement.
>
> The question is asked in one of the notes handed to me, what is the attitude of the Central Committee of the Party to my report? I answer: The Central Committee of the Party examined my report and approved it.[119]

According to the official transcript, the entire audience rose at once from their seats. The ovation was prolonged and stormy. Once he could speak, Lysenko reiterated his arguments against genetics. Weissmanism was the foundation of genetics, which meant geneticists believed that heredity functioned independent of the "living body" and the "conditions of life." Michurists, on the other hand, did not make this separation: heredity was inherent in every part of the organism. Even the sap exchanged between a tree and the branch grafted onto it was capable of transmitting heredity.[120]

This portrayal of how geneticists and Michurinists understood the relationship between the organism and its environment, was reflected in Lysenko's description of their respective attitudes towards the world outside the laboratory. For the Michurinist, the fields of the collective farm—the conditions of life for the Soviet citizen—were the laboratory.[121] Geneticists, meanwhile, isolated behind the ivory walls of elitism, disdained utility.

> It is significant that abroad, in the United States for example, which is the home of Morganism and where it is so highly extolled as a theory, this teaching, because of its inadequacy, has no room in practical farming. Morganism as a theory is being developed *per se*, while practical farmers go their own way.[122]

---

[117] Krementsov, *Stalinist Science*, pp. 173–4. Krementsov also shows that Lysenko's declaration of support from the Central Committee was probably not planned in advance. Rather it was prompted by a question from Zavadovsky, on August 4, concerning the position of the Central Committee on the proceedings (Krementsov, *Stalinist Science*, pp. 172–3).

[118] Krementsov, *Stalinist Science*, p. 174.

[119] *The Situation in Biological Science*, p. 605; Krementsov, *Stalinist Science*, p. 173.

[120] *The Situation in Biological Science*, p. 605–9.

[121] Ibid, p. 616.

[122] Ibid, pp. 613–4.

Lysenko said this same detachment also underpinned the genetic conception of heredity. Mutation was "chance," an unreliable, chaotic process disconnected with reality: "*In general, living nature appears to the Morganists as a medley of fortuitous, isolated phenomena, without any necessary connections and subject to no laws*. Chance reigns supreme."[123] Because this type of "science" did not permit rational planning, it was not suitable for a socialist society. In conclusion Lysenko thanked the "Soviet system" for creating the conditions for the discovery of Michurin, and reserved his final praise for Stalin:

> Glory to the great friend and protagonist of science, our leader and teacher, Comrade Stalin! (All rise. Prolonged applause.)[124]

*Pravda* had begun covering the conference on the fifth day, and continued to publish texts of the presentations for several days after the conference had ended. The attention given to Lysenko's abolishment of genetics was on a scale usually reserved for sessions of the Supreme Soviet.[125] The day he declared that the Central Committee had seen his report and approved it, *Pravda* published a letter of apology from Lysenko's opponent, Iurii Zhdanov. But the letter was not addressed to Lysenko, it was addressed to Stalin.

Zhdanov wrote that his earlier attack on Lysenko had been a mistake, resulting from a "university habit" of expressing his own point of view. His behavior had been "professorial," and he had shown poor judgment in not subjecting Mendelism-Morganism to merciless criticism. Science was no place for reconciliation, Zhdanov said, it was a place where ideas conquered. He was inexperienced, he was immature, and he had made an error in criticizing Lysenko.[126]

Zhdanov's letter, combined with Lysenko's declaration and the ovation which followed, seem to have created panic among the geneticists in the hall. Zhukovsky immediately requested to speak.

> Comrades, late yesterday evening I decided to make this statement. I say late yesterday evening deliberately, because I did not know then of the letter of Yuri Zhdanov which appeared in Pravda today. There is therefore no connection between my present statement and Yuri Zhdanov's letter. I think Vice Minister of Agriculture Lobanov will bear this out, since I phoned him yesterday evening and requested permission to make a statement at today's meeting of the session.[127]

Zhukovsky said that he had barely slept the previous night. Yesterday and today's events, he said, had been "historic" and "profound." As he spoke, his words conveyed increasing concern: "I am a man of responsibility … I am sure that, knowing me, you will believe me when I say that it is not from cowardice that I make this

[123] Ibid, p. 614.

[124] Ibid, p. 617.

[125] Krementsov, *Stalinist Science*, p. 174.

[126] "Yuri Zhdanov's Letter to Stalin," *Soviet Studies* 1, no. 2 (1949): 175–7.

[127] *The Situation in Biological Science*, p. 618.

statement": "Let the past which divided me from T.D. Lysenko (although not always, it is true) be forgotten."[128]

Zhukovsky was followed by Alikhanian, who also began by trying to avoid appearing fearful rather than sincere.

> Comrades, it was not because I had read Yuri Andreyevich Zhdanov's statement in today's Pravda that I requested the chairman to allow me the floor. I decided yesterday to make a statement, and Vice Minister of Agriculture P.P. Lobanov can confirm that I spoke to him on the subject yesterday, August 6.[129]

He too claimed to have been profoundly moved by what had taken place in the past few days. Alikhanian apologized for having allowed his "personal views" to obstruct scientific progress, and promised the "first thing" he would do once the session was over was to not only rethink his attitude towards Michurism, but try to determine what had ever led him to oppose it.[130]

Zhebrak's recantation appeared in *Pravda* a week later. Zhebrak said that as long as "both courses in Soviet genetics" had been supported by the Party he was willing to express his views. Now, however, he realized it was no longer possible to remain in error. He then not only reiterated his agreement with Lysenko, but cited examples in his own past work where their views had aligned. It was the duty of Soviet scientists, he said, to help build socialism and rid the country of every trace of capitalism.[131] Zhebrak demonstrated that he too was ready to change in response to the new conditions Lysenko had created.

Muller drafted a letter of resignation, dated September 24, 1948, and sent it to the Academy of Sciences. Haldane—despite the fact that it was Vavilov who had made him a member—did not.[132] In his letter Muller wrote that he had considered his membership an honor. Though it had been almost a decade since he had heard anything from them Muller assumed, "I am still on your rolls." The deep esteem he felt for the Academy made it all the more painful for Muller to inform them he now found it necessary to sever his connection. Muller said Lysenko was a "charlatan" the Academy had "stooped" to take in. In Nazi Germany genetics had been perverted, and the same thing was now happening in the Soviet Union.

Muller repeated the argument he had made against Lysenko over a decade before: If evolution truly were simply a process of environmental adaptation then the poor and impoverished would be rendered permanently inferior.[133] How could the Soviet Union—founded as a workers' state and on belief in equality—accept such a dangerous, fascist doctrine? No self-respecting scientist could remain a member of the Russian Academy of Sciences now.[134] It is clear from Muller's letter that he believed

[128] Ibid, pp. 168–619.

[129] Ibid, p. 620.

[130] Ibid, pp. 620–1.

[131] Zirkle, *Death of a Science in Russia*, pp. 277–9.

[132] Adams, "Last Judgment," p. 460.

[133] This argument, as noted above, originated with Filipchenko.

[134] Ibid, pp. 307–9.

the outcome of the VASKhNIL conference had great implications not just for Soviet geneticists, but scientists everywhere. He concluded by saying: "The importance of the matters here at issue—including that of the authoritarian control of science by politicians—is in my opinion so profound that I am making his letter public."

Muller's rebuke was widely published in the American press. The Russian Academy's reply to Muller appeared in *Pravda* on December 14, 1948. Any attempt to apply biological laws to social development lowered mankind to the level of beasts, they wrote. Muller had become a propagandist of man-breeding, allying himself with avowed racists and reactionaries. Science divorced from politics did not exist—the only question was whose interests did science serve—the peoples' or the exploiters'? In the United States science has become militarized, serving the goals of profit, oppression and imperialism. What does Muller have to say about that?[135] The response concluded,

> Professor Muller was once known as a progressive scientist. This is a very uncomfortable position in present-day America. Having come out against the Soviet Union and its science, Muller has won the enthusiasm and recognition of all the reactionary forces of the United States.
>
> The Academy of Sciences, USSR, without any feelings of regret parts with its former member, who betrayed the interest of real science and openly passed over into the camp of the enemies of progress and science, peace and democracy.
>
> PRAESIDIUM OF THE ACADEMY OF SCIENCES, USSR[136]

Lysenko had won, but it was the Central Committee of the Communist Party of the Soviet Union which now assumed ultimate authority in all scientific questions.[137]

***

Look back at the quote from *Land in Bloom* that began this chapter. How is Muller's work with radiation characterized? What other postwar fears/concerns/paranoia's does the description remind you of? As for the study conducted by the Atomic Bomb Causality Commission, is it ethical to use food ration cards to attract participants for a study on the impact of radiation upon unborn children?

In Hudson and Richens' list of four ways in which alogical discourse could be enlisted, number four was the insistence that science must be practical. Can you think of contemporary examples where "practicality" is set as the measure of worth? In what instances is this good, in which is it bad?

What about their claim that "achieving that comprehensive understanding of genetical questions" was "the aim of both bodies of investigators?" Was that an accurate assessment of relations at that moment in the controversy?

Compare Dunn and Dobzhansky's two 1946 publications—*Heredity, Race and Society* and *Heredity and Its Variability*. How was the purpose of these publications

[135] Ibid, pp. 309–12.

[136] Ibid, p. 312.

[137] Krementsov, *Stalinist Science*, p. 178.

similar and different? As for the latter, do you think Dunn and Dobzhansky behaved correctly? What do you think of Stadler's criticism? Also, how were Dunn and Dobzhansky's goals similar to, or different from, Hudson and Richens?

Speaking of, do you agree or disagree with Dobzhansky's view that: "The progress of science would be seriously disorganized if all scientists interrupt their work every time somebody publishes a dubious claim. Such claims are disposed of in due course?"

As for Dunn's review in *Science*, was he obligated to tell the editors he had played a role in the translation? What are, or should be, the ethics of reviewing scientific work? Are, or should, the standards be different than in other disciplines such as history, philosophy or literature?

What about Dunn's recurrent comparisons between Lysenko's campaign against genetics and the anti-Darwin movement in the U.S.? Do you agree, or disagree, with Selig Hecht's reaction? Does "scientific freedom" mean being "free" to move to elsewhere if you do not like how—or if—evolution is taught where you live?

What did Dunn mean when he told Waldemar Kaempffert that though the translation of *Heredity and Its Variability* could serve the cause of "those who wish to abuse the Soviet Union," it was still worth doing? Do you think he regretted this later?

Speaking of Kaempffert, what do you make of his role as a go-between for geneticists like Dunn and the "ungenetic" readers of the *New York Times*? How would his missives to his readers have read differently if he had been in touch with Hudson and Richens rather than Dunn? How was the public nature of the controversy—playing out amid burgeoning postwar fears of communism—an important part of this story? What would Stadler say about this exchange between Kaempffert and Dunn?

Going back to the question about what makes Lamarckism appealing raised at the end of Chap. 1, what do you make of Dunn's statement that many people would be inclined to "grasp any straw that seems to confirm their almost innate desire to have evolution interpreted in this way?" Do you think he was right? Do you think this contributed to Lysenko's success, particularly after World War II?

What do you make of Haldane's position? He refused to review *Heredity and Its Variability*, but should he have if he claimed to agree with Lysenko? Why do you think Muller did not tell him about his experiences in the USSR? Would Haldane have reacted differently if he had?

What is the difference between Haldane's role and Shaw's? And why was Shaw interested in Lysenko?

This chapter also addresses the widening divide between Dobzhansky and Dunn. Why do you think Dobzhansky had such an easy time imagining the advent of a fascist system in the U.S.?

Dobzhansky wrote to Dunn that they shared "general sympathies," but he was impatient with the fact that Dunn was willing to give the Soviet government "the benefit of every doubt." What explains this difference between them?

Referring back to the events leading up to the VASKhNIL conference, why would Lysenko be so concerned that the Ministry of Agriculture planned to introduce a tetraploid variety of *kok-sagyz*? How did this threaten him? What did it imply?

In *Land in Bloom*, Safanov wrote,

> The books of the classics of science seem to us to bear an air of majestic calm. Their serenity is deceptive, however. They were intended for battle, and it was in fierce conflict that they won their right to live. All of them rose in rebellion against what in their time was regarded as incontrovertible knowledge. And to this knowledge, that was armed with the power of long tradition, they opposed their rebellious controversial and unprecedented arguments.[138]

How is this relevant to what took place at the VASKhNIL conference? To what extent is Safanov's notion that progress in science is a process of the rebellion of one generation against another correct? To what extent does it, ironically, sound mapped out onto the exact "perversions" Lysenko said were evident in the genetic version of Natural Selection?

Going back over the quotes from the session, how would you describe the strategy of Lysenko and his allies? In what ways are they specific to the context (i.e., Soviet socialism, the Cold War etc.), and in what ways are they examples of scientists and science are often criticized?

Do you agree with the idea that science cannot be divorced from politics? Is it correct—as Lysenko and his allies implied—that to claim so is a just a way to avoid responsibility?

In the aftermath of the VASKhNIL conference many commentators would make comparisons to other instances when scientists (e.g., Galileo) had been forced to "recant." How is the Lysenko case similar or different? Also, what is the appeal of this scenario? How is it related to other examples where you have an individual, or a minority, who are forced to succumb to the will of the majority?

In his book, *Hen's Teeth and Horse's Toes*, biologist and historian of science Steven Jay Gould wrote that Lysenko's declaration—"The Central Committee of the Party has examined my report and approved it"—was "the most chilling passage in all the literature of twentieth century science."[139] Gould's assessment has been shared by many others, but is it correct? What details make the VASKhNIL conference exceptional? Are there any other examples from the history of science (e.g., "Three generations of imbeciles are enough.") that you would consider more shocking?

---

[138] Safonov, *Land in Bloom*, p. 9.

[139] Stephen Jay Gould, *Hen's Teeth and Horse's Toes* (New York: Norton, 1983), p. 135.

# Chapter 4
# Acquired Characteristics

*Because, granted that some bodily defects are transmitted by way of origin from parent to child, and granted that even some defects of the soul are transmitted in consequence, on account of a defect in the bodily habit, as in the case of idiots begetting idiots; nevertheless the fact of having a defect, by way of origin seems to exclude the notion of guilt, which is essentially something voluntary.*

—St. Thomas Aquinas on the inheritance of acquired characteristics, 1256[1]

## 4.1 The Reaction to VASKhNIL

When Lysenko gave his first lecture after the VASKhNIL session the streets surrounding the Academy of Agricultural Sciences in Moscow were jammed with the private cars of officials from the Ministry of Agriculture. Inside the staff was in attendance, and students crowded the hallways to listen in on loudspeakers. A brass band summoned especially for the occasion pumped out a March, and Lysenko proceeded to the podium through cheering throngs. As he spotted older scientists in the crowd he shouted—"Aha! You came to relearn?"[2]

Once his lecture began, Lysenko explained how a horse is only alive through its interaction with the environment. It ingests grass, it breaths air. Once it ceases to participate with the world around it becomes a cadaver: Living bodies always want to eat. Lysenko also wrote that the same elements which are essential for life—air, water, warmth—will cause a dead body to decompose. To continue

[1] Conway Zirkle, "Further Notes on Pangenesis and the Inheritance of Acquired Characters," *The American Naturalist* 70, no. 731 (1936): 529–46, 531.

[2] Medvedev, *Rise and Fall of T.D. Lysenko*, p. 131.

W. deJong-Lambert, *The Cold War Politics of Genetic Research: An Introduction to the Lysenko Affair*, Archimedes 24, DOI 10.1007/978-94-007-2840-0_4,

existing living organisms must consume dead matter, and transform it into energy for themselves.[3]

Dobzhansky was in Brazil, where the U.S. Government had first sent him in 1943 as part of Roosevelt's Good Neighbor Policy, and heard what had happened through clippings sent to him from *The New York Times*.[4] He wrote to Dunn, "You can imagine what I feel."

> Now, should anything be done by American scientists? To keep the record straight if nothing else. To use the biblical expression, if we do not speak out then stones shall speak! Now, as usual, the "speaking" should if possible, come from those who are not habitually engaged in red-baiting. Can AAAS, or the Genetics Society, or the Academy, or I do not know what, make a statement about this terrible business? I am lost to invent what to do, especially while sitting here, so far from the center of things. Pray, tell me what you think? You would be the ideal man to get such a thing started.[5]

Dobzhansky also received clippings from *Pravda*. In one he was described by the Soviet Minister of Education as a "*mrakobes*." Dobzhansky wrote to Dunn:

> The word "mrakobes" is difficult to translate; literally it means "devils of the darkness", and it usually refers to obscurantist fanatics or religious quacks. Anyhow, I can be satisfied that such highly placed authorities seem to be concerned about my insignificant personality.[6]

Meanwhile, Sinnott and Dunn's textbook, *Principles of Genetics*, was withdrawn from circulation.[7]

The first headline referring to Lysenko's victory in *The New York Times* read, "*Pravda* Hits Trend of Soviet Biologists." The story said that the report in *Pravda* named names, but did not specify exactly what the objectionable "trend" was. There was a lot about "slavishly kowtowing to reactionary bourgeois science," and reference to a "theory of a special hereditary substance" which "disarms our scientists and practical workers in their struggle to change the nature of plants and animals," but the exact issue was left unclear. *The Times* elucidated by informing readers that Trofim Lysenko, President of the Lenin Academy of Agricultural Science, claims he can change heredity by altering the environment.[8]

---

[3] T. Lysenko, *Organizm i środowisko* (Warszawa Państwowy Instytut Wydawnictw Rolniczych, 1950), p. 3. See also deJong-Lambert, "The Uses of the Dead in the Science of Life."

[4] Coe, in *The Evolution of Theodosius Dobzhansky*, ed. Adams, p. 24; The Good Neighbor Policy was established by U.S. President Franklin D. Roosevelt to improve relations with Central and South America.

[5] Correspondence, Th. Dobzhansky to L.C. Dunn, September 9, 1948. B: D917 L.C. Dunn Papers. Dobzhansky, Theodosius 1948–1949. The American Philosophical Society.

[6] Undated correspondence, Th. Dobzhansky to L.C. Dunn. B: D917 L.C. Dunn Papers. Dobzhansky, Theodosius 1948–1949. The American Philosophical Society.

[7] The Reminiscences of L.C. Dunn, 256. Oral History Research Office. Columbia University, 1961.

[8] "Pravda Hits Trend of Soviet Biologists," *New York Times*, August 13, 1948.

A follow-up story was more specific: "Lysenko Crushes Geneticists in Russia; Gets Party Backing for His Theories." "After more than a decade of struggle," it read, Lysenko had emerged victorious. The genetic theories of Mendel and Morgan were now banned from Russian laboratories, textbooks and university courses.[9]

Another article in the *Times* referred to a debate between Lysenko and Vavilov in 1936, sponsored by a "party organ, *Under the Red Banner*," as the origin of the controversy, conflating the discussion Muller had participated in on "issues in genetics," with the debate between Vavilov and Lysenko, sponsored by *Under the Banner of Marxism*, which had taken place in 1939. In any case, the *Times* claimed that this had been the beginning of the end for Vavilov: "He died, a broken victim of Soviet tyranny in circumstances that have never been officially explained." The story also pointed out that what had taken place in Moscow was the same as if Americans had to accept whatever the Republicans or Democrats thought about science, depending upon which party was in power. With the Soviets demanding greater portions of Germany, claiming that about every major scientific and technological breakthrough had begun in Russia, and charging composers like Shostakovich with "propagation of musical heresies," Lysenko—according to the *Times*—"has a good chance of winning his point."[10] A story published a few days later said it was ironic that, given Vavilov's fate, he was the one first responsible for making Lysenko's name known outside of the Soviet Union. Zhebrak's criticism of Lysenko and subsequent recantation were also noted.[11]

One reader, John H. Vincent, wrote to the editors claiming that while the fate of Vavilov "at the hands of the ineffable Lysenko is deplorable," it was as hard to be a Lamarckist in the United States as it was to be a Mendelist-Morganist in the Soviet Union. Vincent said advocating the belief that acquired characteristics could be inherited at best meant the end of an academic career, and at worst was taken as a sign of mental illness. Any professor who believed it might not be "sent to Alaska," but "he might soon find himself running a filling station." On the other hand, Vincent asserted, the position of geneticists is firmer in direct proportion to their evidence becoming less secure. Evolution is so slow even a million years of lab tests might not be able to prove Lamarckism to the satisfaction of geneticists. According to Vincent, geneticists believed the "changes which resulted in the formation of the millions of species of animals and plants were due to accident, cosmic ray or simply unknown causes." Who knows, he wrote, maybe Lysenko's on the right track?[12]

9 "Lysenko Crushes Geneticists in Russia," *New York Times*, August 19, 1948.

10 "Lysenko Again to the Fore," *New York Times*, August 22, 1948.

11 "Russian Recants Bourgeois Science; Accepts Party Line in Genetics," *New York Times*, August 25, 1948.

12 "Letters to the Times," *New York Times*, August 27, 1948.

Press coverage continued to inform the American public of genetic research institutes being closed down in the Soviet Union, and geneticists who were forced to recant or lose their positions. The *Wall Street Journal* published a synopsis of events, declaring sarcastically that—"Bourgeois science has had its comeuppance."[13] The *Hartford Courant* provided a line—"That whirring sound is Gregor Mendel stirring in his restless grave"—and asked, "…what can there be of intellectual freedom when a Soviet stooge can become a Joshua and order the sun of scientific progress to stand still?"[14]

*The Washington Post* adopted a similarly mocking tone:

> It now occurs to us that the profound mystery of the Russian Communist mind, which has been a source of so much bafflement to the statesmen and journalists of the Western world, might be cleared by reference to the science of epistemology… Politics have become for the devout Communist a religious absolute to which all questions of ethics, aesthetics, natural science and even of economics and history must be subordinate. … Anyway, it is only by recognizing the religious character of communism and the constitution of the Communist Party as a universal and indivisible church, that it will ever be possible to understand the Communist mind.[15]

Like the press, politicians who discussed Lysenko were also attracted to analogies between the "Communist mind" and religious faith. Assistant Secretary of State, George V. Allen, expressed his disgust at Lysenko's victory in a speech before the U.S. Advisory Commission on Educational Exchange. The closest we have come to such a thing in the United States, Allen said, was the attempt to outlaw the teaching of Darwinian evolution in Tennessee: "It is hoped there will be not more such monkeyshines."[16]

Comparisons between Lysenko and fundamentalist anti-evolutionists led to the contention that he was indicative of a more general "backwardness" in Russian science. The spread of the Lysenkoist doctrine to medical biology and the threat posed to research in antibiotics inspired the comforting speculation that at this rate, the Soviets would never develop the atomic bomb. *The New York Times* also reported on radio broadcasts from Moscow to Soviet secondary school teachers mandating that they conform to Marxist-Leninist teachings in biology.[17] A columnist at the *Times* also claimed that Lysenko's doctrine was like believing a baseball player's son would automatically throw a ball as well as his father.[18] *The Washington Post*, *Hartford Courant* and *L.A. Times* all reported a rumor circulating in American scientific circles that someone visiting one of Lysenko's greenhouses had pulled a

---

[13] "'True' Science," *The Wall Street Journal*, August 20, 1948.

[14] "Repeal of Mendel," *The Hartford Courant*, September 1, 1948.

[15] "The Mind of the Kremlin," *Washington Post*, August 22, 1948.

[16] "No Walls Between Students," *New York Times*, September 12, 1948.

[17] "Biology in Soviet Must Be Partisan," *New York Times*, September 5, 1948.

[18] "The Word: Russian Science," *New York Times*, August 29, 1948; "Soviet Widening Scientists Purge," *New York Times*, September 23, 1948.

giant tomato off a vine and discovered it was made of wax.[19] *Time* magazine informed readers that: "Henceforth, all vegetables, flowers and other plants in the U.S.S.R. will grow straight along the Marxian line."[20]

## 4.2 Scientists, Intellectuals and the Cold War

August 25–28, 1948 the World Congress of Peace, a gathering of 500 prominent representatives of science, literature and art from East and West, was held in Wrocław, Poland. The purpose of the Congress was to see if writers, artists, scientists and academics could locate the common ground that seemed increasingly elusive to statesman. Julian Huxley co-chaired as president of UNESCO, and J.B.S Haldane also attended. The Congress was controversial from the start, used by the delegates of East Bloc nations to condemn Western culture as a bourgeois barbarism of Frigidaires and automobiles.

The congress was also the first attempt to extend the attack on genetics beyond the border of the Soviet Union.[21] The Soviet delegations failed in their effort to have a rejection of genetics included in the final resolution, but Huxley refused to sign it anyway. Huxley declared that dialogue and discussion had been replaced by a useless process of "issuing manifestos to each other," and dismissed the proceedings as "tendentious and unfortunate."[22] Haldane, on the other hand, considered the Congress a momentous success. He called on scientists to unite with the progressive powers of Eastern Europe, and said American bacteriological warfare researchers were, "just as much traitors to the human race as were the Nazi doctors in Auschwitz."[23]

Huxley began to use his position with the UN as a platform to publicize the situation in Soviet genetics. On the eve of a UNESCO session in Beirut, Lebanon, Huxley said that during his recent official visits in Yugoslavia, Poland, Hungary and Czechoslovakia, he had seen immense efforts being made in science, culture, and education. Huxley said he felt no barrier between himself and his academic colleagues in the East, but rather a great willingness to cooperate and exchange

[19] "Matter of Fact: Phony Tomato," *The Hartford Courant*, August 23, 1948; "The Phony Tomato," *The Washington Post*, August 26, 1948; "Marxism as Applied to Growing Tomatoes," *Los Angeles Times*, August 25, 1948.

[20] *Time*, September 6, 1948, p. 66.

[21] The Wrocław congress was also the first time French and Italian biologists learned what had happened at the VASKhNIL session. Francesco Cassatta, "The Italian Communist Party and the 'Lysenko affair' (1948–1955)," *Journal of the History of Biology* 45, no. 1 (2012).

[22] "Huxley Asks End to Attacks," *New York Times*, August 28, 1948; "Huxley Denounces Breslau Conference," *New York Times*, September 2, 1948.

[23] "Haldane Hits U.S. Doctors," *New York Times*, September 6, 1948; "Briton Assails Intellectual 'Parley,' for Teaching War Not Peace," *New York Times*, August 27, 1948.

information: "Indeed, so far as I could ascertain, barriers to cultural exchange are at least as great between these countries and the USSR, as they are between them and Western Europe."[24] Huxley also said he saw no purpose in seeking to unify the differing ideologies of East and West, and cited the "fantastic" theories of T.D. Lysenko as part of the reason why. The rationalist methods, traditional in the West, were incompatible in a place where the "primary canons of scientific method are no longer recognized."[25]

The annual general conference of UNESCO (an organization the Soviet Union continued to boycott) was held in Beirut in November, 1948, just as the Arab-Jewish conflict was boiling over. An illegal communist demonstration was broken up with gunfire outside the plenary session, and the Lebanese delegate led a protest of Arab states against "world Jewry" and Zionism. Huxley, in his position as director of UNESCO, once again took the opportunity to criticize Lysenko, declaring in his annual report: "I have heard with my own ears Lysenko lead the attack on 'bourgeois' genetics, which has now resulted in the dismissal of some of the leading biologists in the USSR and the straitjacketing and distorting of one of the most important branches of science."[26]

Muller, meanwhile, attacked Lysenko in the popular press. Four months after the VASKhNIL conference Muller published two articles—"The Destruction of Science in the USSR" and "Back to Barbarism Scientifically"—in the *Saturday Review of Literature.*[27] In the first he wrote that as far Soviet genetics was concerned, "all that we can now hope to do is to conduct an autopsy."[28] Muller described the "success" of Lysenko's work as "dubious," and said it gave "him no more claim to being a geneticist than does the treatment of dogs for worms." "Lysenko's writings along theoretical lines," he added, "are the merest drivel."[29] Muller also called "Lysenkoism" a "dangerous superstition," comparable to the "belief that the earth is flat," and intended to "degrade rather than advance humanity."[30]

In "Back to Barbarism Scientifically" Muller also described elements of Lysenkoism he saw present in American society. The article included a photograph of William Jennings Bryan and Clarence Darrow, captioned: "When we criticize the Soviet attack on science, let us not forget…the assault on the teaching of evolution during the Scopes trial in Tennessee, led by the politician William Jennings Bryan."

---

[24] "Cultural Deviation in Soviet Bloc Seen," *New York Times*, November 17, 1948.

[25] Ibid.

[26] "U.S. Aide Assails Soviet at UNESCO," *New York Times*, November 20, 1948; "Abroad," *New York Times*, November 20, 1948.

[27] H.J. Muller, "The Destruction of Science in the USSR," *The Saturday Review of Literature*, December 4, 1948, pp. 13–5, 63–5; H.J. Muller, "Back to Barbarism Scientifically," *The Saturday Review of Literature*, December 11, 1948, pp. 8–10.

[28] Muller, "The Destruction of Science in the USSR," p. 13.

[29] Ibid.

[30] Ibid, p. 65.

Muller said that the Scopes trial was "only the most publicized" of many similar "scandals"[31]:

> The writer well recalls a session of the Texas legislature at which a preacher by the name of Norris delivered, by special invitation, a fanatical two-hour harangue on the doctrine of biological evolution and its "dangers"—Bolshevism, "nigger-loving," and the anti-Christ. The legislature listened attentively and frequently applauded. No qualified person was allowed to state the case for science. Subsequently the lower house passed a bill forbidding the teaching of evolution in elementary and high schools, and the state textbook commission ordered the removal of all mention of the subject from school textbooks, an order that was rigorously executed.[32]

In addition to the threat to science posed by anti-evolutionists, Muller also pointed out the "danger" created by the dependence upon private foundations to fund scientific research. "Why," he asked, "should the scientist have less prestige than the businessman and be considered less qualified for handling funds in his own field?" The "gravest present danger," however, Muller said, stemmed from the "activities of super-patriots," such as the House Un-American Activities Committee (HUAC), "who, on the plea that they are battling totalitarianism and defending democratic freedoms, are themselves attempting to fasten the very evils they warn against upon our own country." According to Muller, when criticizing the "shocking treatment accorded scientists in Nazi Germany, and which is now being given them in the USSR, we must also exert ourselves to prevent the same thing from happening in our own midst."[33]

Of the six letters published in response to Muller's articles, only one was positive. Muller was accused of being "unscientific" and "emotional." One reader, referring to his "vehemence," accused him of not offering any conclusive evidence for why he disagreed with Lysenko, and another said what Muller had written was a "political diatribe." Another respondent wondered why Muller criticized Lysenko, since U.S. citizens should "welcome anything that decreases" the "ability" of the Soviet Union "to conquer us."[34]

Best-selling author Howard Fast was also critical of Muller.[35] Fast accused Muller of conducting a "frenzy of unrestrained name-calling," and characterized Lysenko's arguments as "calm and reasoned" scholarly fact gathering. Fast said Muller's misrepresentation made one wonder why the Soviets would "rape science" and "cut their own throats." Fast admitted to having only a layman's

---

[31] Muller, "Back to Barbarism Scientifically," p. 9.

[32] Ibid.

[33] Ibid, 10.

[34] *Saturday Review of Literature*, January 8, 1948, 23–4.

[35] Howard Fast (1914–2003) was well-known novelist and member of the Communist Party USA. In 1950 he was called before the House Un-American Activities Committee and was forced to self-publish his most important novel, *Spartacus*, after being blacklisted by major publishers. The novel was a success, and the production of the film version is seen as an important victory against the anti-communist movement in the United States. Other significant works from this period include an essay, "Why the Fifth Amendment?," where Fast spoke out against McCarthyism.

knowledge of biology, but said "if Lysenko were as wrong as Prof. Muller considers him to be, he could best be demolished through a careful and intelligent refutation of his arguments." Muller's "hysteria" made the "judicious reader doubt the ground he stands on."[36]

## 4.3 Myth Making, Propaganda and Purging

As the Stalin Plan for the Transformation of Nature got underway in the Soviet Union, posters of Stalin dressed as a medieval knight appeared in shop windows with the slogan: "We will conquer drought too!" Tree belts were planted across an area equivalent in size to Great Britain, France, Italy, Belgium and the Netherlands combined to alter the climate in order to provide fresh fruit year-round for Soviet children, and feed their ancestors for centuries to come. The trees were mostly oak, planted 30 m thick. If lined in a row they would circle the equator 50 times. They were placed in clusters, as Lysenko recommended, on the principal of "self-immolation"—that the weaker trees would sacrifice themselves for the stronger. As Lysenko explained: "Even without reference to biological theory, it can be proven in purely practical terms that, if one thing hinders two things, these two can always be united, even if temporarily, against their common enemy."[37]

Soviet filmmaker Alexander Dovzhenko's above-mentioned biopic of Michurin was widely advertised across the Soviet Union and later released in the U.S. under the title "Life in Bloom." It was one of the first color films ever made in the USSR, as well as one of the first to be promoted as a preview shown before other films.[38] Lysenko himself served as an advisor on the set, assuring the story was told correctly.[39] A review published in *The New York Times* was dismissive, and noted that, "the work of heredity and environment of his scientific heir, Prof. Trofim Lysenko, which recently officially replaced that of Mendel and Nobel laureate Thomas Hunt Morgan in Soviet texts, is not specifically mentioned."[40]

---

[36] *The Journal of Heredity* 40, no. 7 (1949): 194.

[37] Soyfer, *Lysenko and the Tragedy of Soviet Science*, pp. 205–7. See also Stephen Brain, "Stalin's Environmentalism," *Russian Review* 69, no. 1 (January 2010).

[38] *Journal of Heredity* 40, no. 7 (1949).

[39] "Soviet Film Hails Michurin, Plant Breeder, Whose Theories Stirred Up Heredity Battle," *New York Times*, January 3, 1949.

[40] "Movie Review, Michurin (1948), The Screen in Review; 'Life in Bloom,' Film of Life of Russian Scientist, Is New Picture at the Stanley Theater," *New York Times*, May 9, 1949; For further information on the film see deJong-Lambert and Krementsov, "On Labels and Issues."

Meanwhile, the purge of genetics spread across the Communist Bloc. In Kiev a Moscow delegation arrived to report on the new official doctrine in biology, and geneticists were led across a stage one by one in front of interrogators.[41] In Czechoslovakia the President of the Czechoslovak Agricultural Academy in Prague ordered the elimination of "impotent and sterile idealistic biology of the laboratories which lacks all trace of reality."[42] At a conference held in Bulgaria in April, 600 participants witnessed promises to master the new ideas coming out of the Soviet Union.[43] In Romania an ophthalmologist who had saved the now-deposed Queen Helen from total blindness was denounced for scientific deviation. As a pioneer of corneal transplants he had failed, according to the Chairman of the Romanian Academy of Sciences, to acknowledge the superiority of Marxism-Leninism.[44]

The reactions in Western Europe were even more curious. West German geneticist Hans Nachtsheim argued that what had taken place at the VASKhNIL conference was even worse than the collaboration of German geneticists with the Nazi regime. According to Nachtsheim, the German government was interested in science because science is useful. Is that scientists fault? In the case of Lysenko, however, the situation was clearly the opposite. This time the government, not the scientists, was in control.[45]

In Italy the controversy was framed by the country's ambiguous position between the poles of Cold War politics. Italian geneticists were interested in what Lysenko had to say, but the line drawn by Stalin at the 1947 conference of the Cominform in *Szklarska Poręba*, Poland—you are either for us or against us—made their situation impossible. As a result, despite the sympathy of some Italian biologists, no Italian translation of the VASKhNIL session was ever published.[46]

Delegations of scientists also made pilgrimages to Moscow to meet Lysenko, with a stop-over in Michurinsk to view the model farm. A Polish scientist recalled Lysenko as spellbinding, and the orchards of Michurinsk as immaculate as an assembly line, displaying an unnatural order. Other biologists compared Lysenko to Savonarola—a comparison which was also becoming increasingly popular in the West. Lysenko's lab was said to be populated by rows of experimenters perched on stools in white lab coats, examining microscopes in which nothing could be seen. When he spoke his mouth frothed and he challenged opponents, though no one dared to contradict him.[47]

---

[41] Soyfer, *Lysenko and the Tragedy of Soviet Science*, p. 193.

[42] "Czechs 'Revise' Biology," *New York Times*, March 6, 1949.

[43] B: Z67 Conway Zirkle Papers, Dimitrova, Elena--'Lysenko in Bulgaria' (from Free Bulgaria 1949, p. 244). The American Philosophical Society.

[44] "Rumania Reported Fighting Partisans," *New York Times*, July 25, 1949.

[45] Alexander von Schwerin, "Lysenkoism and the Reform of Postwar West German Genetics." Paper presented at the International Workshop on Lysenkoism, December 5, 2009.

[46] Francesco Cassata, "The Italian Communist Party and the 'Lysenko affair'."

[47] Gajewski, "Lysenkoism in Poland," pp. 426–7; Szczepan Pieniążek, *Dookoła Sadowniczego Świata* (Warszawa, Poland: Państwowe Wydawnictwo Rolnicze i Leśne, 1965), pp. 450–2; Leszek Kuźnicki, Conversation with author, 11 August 2003.

Among the most notorious of Lysenko's followers was Olga Lepeshinskaya. Lepeshinskaya had known Lenin while he was still in exile in Geneva before the revolution. In the 1930s she claimed to have produced cells in egg-yolk, thus refuting Rudolf Virchow's work showing that cells only originate from other cells. Her "findings" were ignored. However she was able to use her Party connections to receive support from Stalin.[48]

After the war Lepeshinskaya published a book. It was nominated for a Stalin Prize but only Lysenko voted for it. Lepeshinskaya was so insulted she launched a campaign against, "Virchowism," thus terrorizing her enemies. Lepeshinskaya's experiments were now widely confirmed, as microscopes began to miraculously display her imaginary living matter throughout the USSR. One admirer offered up evidence of living bone growing in a jar of dead worms. These unlikely results were justified by a quote from Stalin: "Facts are stubborn things." Meanwhile, Lepeshinskaya's living matter was also observed growing in abdominal cavities sewn with cereal grains. Another method for creating life, reported in a Ph.D. dissertation, was to inject a powder made from ground pearl buttons into animals. The living matter arose from the powder which, the researcher claimed, was unsurprising given that mother of pearl comes from shell, and shell was once alive.[49]

If Lepeshinskaya's work was supported for political reasons, the theories of Ivan Pavlov were adopted for the purpose of credibility. Pavlov's experiments in physiology were famous—the phrase "Pavlov's dogs" recognized worldwide as evidence of involuntary, conditioned behavior. The idea of taming to conformity (e.g., "mentor method") was also useful to the Lysenkoist doctrine of training and transforming nature. The key question, however, was inheritance.

Pavlov, like many other physiologists in the early twentieth century, had endorsed the idea that acquired behavioral traits could be inherited. One of his colleagues tested the idea with mice who ran a maze to find food on a feeding rack after a bell was sounded. The experimenter claimed each generation of mice subjected to the test improved on the performance of their forbears. Pavlov repeated his claims on a lecture tour of Britain and the United States. In Britain the reception was enthusiastic, but in the United States T.H. Morgan attended Pavlov's talk and raised numerous objections. Ultimately Pavlov decided the results had more to do with the enthusiasm of the scientist than the actual transmission of bell-conditioned feeding behavior. He became interested in genetics and had a statue of Gregor Mendel erected in front of his laboratory. In 1948 Mendel's statue was removed and put in storage: Pavlov, dead for over a decade, was now a Michurinist.[50]

[48] Medvedev, *Rise and Fall of T.D. Lysenko*, pp. 134–5.

[49] Medvedev, *Rise and Fall of T.D. Lysenko*, pp. 133–4, 182–4; Soyfer, *Lysenko and the Tragedy of Soviet Science*, pp. 213–7.

[50] Krementsov, *Stalinist Science*, pp. 260–74.

Claiming the inheritance of acquired behavioral characteristics aligned Lysenko's doctrine with the Marxist vision of historical progress. The attributes acquired through socialism, inherited by succeeding generations, would lead to the evolution to communism. However, the case of Pavlo D. Lysenko, Trofim D. Lysenko's brother, indicated that the conditions of Soviet socialism were far from beneficial to every scientist.

Pavlo sought asylum in the United States just before his older brother banned genetics in the Soviet Union. According to descriptions published in the *New York Times* and the *Journal of Heredity*, Pavlo was shy, modest, clean cut, and had wavy brown hair. Pavlo had studied coal chemistry at a local university, graduated in 1932, and become part of the research staff at a scientific institute. After only two years, at age 25, he attained the prestigious rank of Scientific Worker. His research testing methods for the conversion of coal into coke were highly successful. Pavlo published widely and aroused jealousy.

When the Second World War began his enemies arranged to have him drafted, even though he was officially exempt as a scientist. Pavlo was captured by the Germans and ended up in Munich by the time the Americans arrived. He supported his family by converting inedible horse chestnuts into artificial honey. Like other "non-returners," he feared being kidnapped and forced to return to the Soviet Union. After appealing to the International Rescue Committee Pavlo was told he had four days to prepare for departure. Pavlo sold his makeshift factory for a tenth of what it was worth. He did not want to miss his chance to leave, and he was never going back to the Soviet Union.[51]

## 4.4 Impregnation by Wind

In 1949 Conway Zirkle edited a book on the Lysenko affair, *Death of a Science in Russia*. Zirkle was a botanist and historian of science at the University of Pennsylvania whose interest in the Lysenko controversy was a product of his concern with Marxism, in addition to a general fascination with the occasional lapses into accuracy of scientific ideas which ultimately proved unsound. Zirkle not only believed that Marxist philosophy had a disastrous impact upon biology in the Soviet Union, but that "Marxian biology" continued to have a pernicious impact upon U.S. academia.[52] Zirkle's efforts against Lysenko were not motivated by concern for Soviet geneticists, but by his desire to use Lysenkoism as a weapon against leftists in the West.

An early interest of Zirkle's was the ancient belief in impregnation by wind. The notion that females of certain species—mares, vultures, hens and women—could be

[51] "Lysenko's Brother Takes Refuge Here," *New York Times*, September 11, 1949; "Lysenko's Brother Escapes to U.S.," *The Journal of Heredity* XL, no. 9 (1949): 251.

[52] See especially Conway Zirkle, *Evolution, Marxian Biology, and the Social Scene* (Philadelphia, PA: University of Pennsylvania Press, 1959).

impregnated by the wind could once be found from Portugal to Japan, and all the countries in between. For a scientific error it had a very honorable pedigree. References can be found in Aristotle, Virgil and Pliny. St. Augustine wrote of "many things which reason cannot account for, and which are nevertheless true."[53] These included salt from Sicily which became fluid when thrown into fire and crackled like fire when thrown into water; a stone in Arcadia called asbestos because once lit could not be put out; the wood of a kind of fig tree found in Egypt which would not float on water; apples of Sodom which when touched cracked and tumbled into ashes; and the mares of Cappadocia which are impregnated by wind.

The legend of vultures being impregnated by the east wind came from Egypt. It survived in Christianity as evidence for the possibility of the virgin birth of Jesus Christ. In the fourth century A.D. St. Ambrose of Milan wrote:

> What say those who are accustomed to smile at our mysteries when they hear that a virgin may generate and do they esteem impossible bearing by an unmarried girl whose modesty no custom of man violates? Is that thing thought impossible in the Mother of God which is not denied to be possible in vultures?[54]

Hens, on the other hand, were supposedly impregnated by the west wind. For centuries their sterile unfertilized eggs were known as "wind-eggs." Reference to wind eggs appears as late as the seventeenth century at the conclusion of a scientific discussion by William Harvey, who first identified blood circulation: "Hence the ancients, when with this wind blowing in the spring season, they saw their hens begin laying, without the concurrence of the cock, conceived that zephyrus, or the west wind, was the author of their fecundity."[55]

The first mention of women being impregnated by wind can be found in Chinese literature, associated with mythical countries populated only by females. Fertilization by the wind enabled them to perpetuate their society. The same idea appears in the Muslim world around 1,000 A.D., where an island located at the edge of the China Sea was purportedly populated by women who bore only females after being impregnated by the wind. The belief survived in Europe into the eighteenth century as part of natural theology, cited by a London clergyman as proof that the souls of children were not generated from the souls of their parents.

Though the possibility of impregnation by wind had long been widely understood to be false, the belief interested Zirkle for two reasons: one, it originated before the discovery of impregnation by wind which does indeed take place in nature—pollination—and two, wind was identified across cultures as the mechanism of impregnation. The fact that pollination existed conceptually—and falsely—prior to its realization as a scientific truth, demonstrated for Zirkle the proximity of rationality and ancient fantasy. As for why wind would be so popularly believed to be the agent of impregnation, Zirkle speculated that to early man there was

---

[53] Conway Zirkle, "Animals Impregnated by Wind," *Isis* 25, no. 1 (1936): 101.

[54] Zirkle, "Animals Impregnated by Wind," p. 107.

[55] Ibid, p. 114.

something immaterial about the wind, which could not be grasped, held or weighed. It blew where it wanted, and though it could be heard no one knew where it came from or where it went. The wind of storms seemed a power to be appeased, but the breeze blowing in and out of the lungs was the breath of life. The answer could also be found in etymology: the Greeks used the word *pneuma* for both air and spirit, the Latin *spiritus* was a breath and a ghost, and the Germanic ghost or *geist* originally meant a breeze—gust. Even as recently as a century ago, Zirkle believed, it would take a very bold philosopher to deny that a *pneuma*, a spirit or a ghost could cause a virgin to bear offspring.[56]

Dobzhansky wrote to Zirkle that he had encountered the same belief, as well as that urinating into the wind causes gonorrhea, while traveling in Turkestan. Dobzhansky noted that the Kazakh and Kirgiz tribesmen were aware of how gonorrhea is actually transmitted, however this did not keep them from believing that wind was a culprit as well. Dobzhansky also pointed out that since their way of life depended almost entirely on stock breeding, especially horses, the locals were familiar with the empirical rules of Mendelian inheritance. They observed inheritance in coat colors, and their language while, "rudimentary in other fields," was rich in terminology referring to domestic animals.[57]

Dunn also recognized that truth and superstition often co-existed quite comfortably, and accepting this was part of what being a scientist was all about. In reply to a letter he received from a correspondent, Miss Deborah Bacon, who solicited his opinion on the Lysenko controversy, Dunn wrote that part of the problem was the impossibility of proving acquired characteristics are *not* inherited. This would require demonstrating a universal negative, and such disproof could only ever approach completeness. There would always exist a finite chance an exception could be found. For this reason, he said, scientists as a rule "are rather tolerant of any heresy."[58]

Dunn wrote an article, which was reprinted in the *Overseas News Agency* under the title, "Scientist Finds Russian Dictum on Genetics Purely Political, But Opposed to Nazi Idea," saying as bad as Lysenko might be, the Nazi belief that "blood was all-powerful" was worse. While there was no experimental justification for Lysenko's notion that environment was all that mattered in shaping evolution, the idea had, "never lacked friends among those in all countries to whom apparent determinism of heredity is philosophically repugnant."[59]

In the *Bulletin of the Atomic Scientists*, a journal founded in recognition of the new political and social responsibilities required by the nuclear age, Dunn wrote

---

[56] Zirkle, "Animals Impregnated by Wind," pp. 95–130.

[57] Correspondence, Th. Dobzhansky to Conway Zirkle, April 23, 1948. B: Z67 Conway Zirkle Papers. Dobzhansky, Theodosius. The American Philosophical Society.

[58] Correspondence, L.C. Dunn to Miss Deborah Bacon, March 21 1949. B: D917 L.C. Dunn Papers. Lysenko Controversy in the U.S. The American Philosophical Society.

[59] "Scientist Finds Russian Dictum on Genetics Purely Political, But Opposed to Nazi Idea," B: Z67 Conway Zirkle Papers. Dunn, L.C. Concerning the Lysenko Report: 4. The American Philosophical Society.

that Lysenko's views were adopted because they freed Soviet agricultural workers from being bound by scientific laws they themselves did not make. It was a moral, not a scientific issue, and thus required arguments that were "hortatory and patriotic" rather than "logical and objective": "To discuss it as an aberration is to abandon hope of understanding how, in a revolutionary state, all parts of life are connected by a political lifeline."[60]

Zirkle made no such equivocations. Observing the recantations of Bulgarian geneticists he commented: "Tyranny compels many queer acts."[61] Zirkle received a letter from a Hungarian biologist, László Havas, pleading for help. Havas' passport had been taken, his laboratory raided, and he had fled to the West. He was penniless and dependent on the kindness of friends like Zirkle.[62] As he wrote:

> I had to confront this dilemma: to submit, against my scientific consciousness and that of a free man, or to resign and risk starving abroad but in liberty. I have chosen the last solution. I did not have to wait very long for results...[63]

Zirkle dismissed Lepeshinskaya's theories as 100 years out of date, and referring to the extension of scientific dictatorship to the ideas of Pavlov, he said: "It is truly incredible but at present the Communist party line both prescribes and proscribes theories of digestion and of functioning of the central nervous system."[64]

## 4.5 Death of a Science in Russia

Zirkle was morbidly fascinated by the intellectual climate—at least as he perceived it to be—in the Soviet Union. In a speech entitled "An Appraisal of Science in the USSR," given in the ballroom of the Philadelphia Municipal Auditorium, he described current developments in Soviet astrobotany:

> The astro-botanists claim to have discovered life on Mars. They also claim that the climate of Mars resembles that found in parts of Siberia, hence the flora of these regions are being studied so that the astro-botanists can learn what plant life is like on Mars. We really cannot exaggerate or parody this stuff.[65]

Zirkle's dismissive attitude towards Soviet science was very apparent in his book, *Death of Science in Russia. Death of a Science in Russia* contained contributions from Muller, Dobzhansky, Dunn and a number of other scientists, explaining the Lysenko affair to an American audience. University of Pennsylvania Press

---

[60] *Bulletin of the Atomic S cientists*, May (1949): 143.

[61] B: Z67 Conway Zirkle Papers, *Death of a Science in Russia* (Preface to Japanese Edition). The American Philosophical Society.

[62] B: Z67 Conway Zirkle Papers. Havas, Laśzló. The American Philosophical Society.

[63] Ibid.

[64] "An Appraisal of Science in the U.S.S.R.": 10. B: Z67 Conway Zirkle Papers 'An Appraisal of Science in the U.S.S.R.'. The American Philosophical Society.

[65] Ibid, p. 16.

published it in hardback for $3.75, and Zirkle wished to compensate the authors for their contributions. The Press had no funds for this purpose, so Zirkle took it upon himself to determine what they were all owed "volumetrically," calculated by the number of pages they had contributed as a percentage of the royalties. He sent contributors a form letter enclosed with checks for various amounts which indicated that the largest contribution was 4.5%, which entitled the author to $23.63. The smallest contribution was .25%, for which the author was paid $1.32. As Zirkle wrote—"This latter amount is, at present, the cost of two cocktails and the editor hopes that the author who receives this check drinks to the confusion of Lysenko twice."[66] Zirkle kept a percentage reflecting his own contribution, and retained as an editorial fee any money which—since the second half of the book was mostly reprints of the official transcript from the 1948 session, *The Situation in Biological Science*—might otherwise go to the Russian government. "This latter," he said, "is an example of depositing the Philistines and is, the editor has been assured, a high moral activity."[67]

In a brief introduction to the excerpts from the official transcript Zirkle confessed that the translation he was using was different. The purpose, he said, was to produce a text that was more "literal," than the one the Soviet government had provided. A second reason, according to Zirkle, was to present Lysenko as he really was: vulgar and uneducated.

> Scholars consulted by the editor have agreed that Lysenko expresses himself in very bad Russian. We have tried to keep Lysenko's flavor (this side of incoherence) if mediocre English can ever be given the flavor of bad Russian. … Our version contains certain crude expressions used by Lysenko which are not in the official translation.[68]

For example, in a line from *The Situation in Biological Science*, quoted above, party philosopher Mark Mitin called the notion of pure hereditary lines "reactionary and preposterous piffle." Zirkle's colleagues translated this to read "reactionary and insane delirium."[69] Whereas in the official translation Lysenko said of Malthus' influence upon Darwin—"Many are still not clear about Darwin's error in transferring into his teaching Malthus' preposterous reactionary ideas on population"—in Zirkle's version, Lysenko said: "Darwin's error of transferring into his own doctrine the mad-brained reactionary Malthusian scheme on populations is up to the present not realized by many."[70]

The editorial hand was also heavy in the use of footnotes. The section quoted above where Prezent described the two-sided museum display, did not appear in the official transcript. Zirkle not only chose to include the anecdote in his excerpts, but

---

[66] Correspondence, Conway Zirkle to contributors, *Death of a Science in Russia*. B: D917 Dunn. Zirkle, Conway. The American Philosophical Society.

[67] Ibid.

[68] Zirkle, *Death of a Science in Russia*, p. 98.

[69] Zirkle, *Death of a Science in Russia*, p. 151; *The Situation in Biological Science*, p. 265.

[70] *The Situation in Biological Science*, p. 13; Zirkle, *Death of a Science in Russia*, p. 100.

added a footnote implying that Prezent was visiting the museum purposely to inspect the exhibition.

> This is a most damning admission for an intelligent man to make. Prezent could hardly have been unaware of what he was doing. His jeer gives a real picture of how scientists have to work in a totalitarian state and what expedients they have to adopt to keep their work going. In exposing these conditions, Prezent has performed a valuable service to world science.[71]

Another section quoted a participant who claimed that early geneticists' belief that the gene was inalterable was proof that "the theory of development is distasteful to them." Zirkle invoked Nazi science and quoted Lewis Carroll.

> This falsehood is repeated frequently by the Party liners, and mere repetition, as Goebbels knew, convinces a certain common type of mind. On the other hand, Lewis Carroll satirized such standards in *The Hunting of the Snark*, thus:
>
> "Just the place for a Snark! I have said it twice:
> That alone should encourage the crew.
> Just the place for a Snark! I have said it thrice:
> What I tell you three times is true."[72]

As Dobzhansky had said, the purpose of translating Lysenko was to force him to "stand on his own two feet."[73] It is also clear that those doing the work of translation were determining Lysenko's posture. In Zirkle's case the intent was to make not just Lysenko, but Soviet science in general, appear reactionary and ridiculous.

Zirkle's motives for engaging in the Lysenko controversy, like the outcome of the VASKhNIL conference itself, were a product of the Cold War. The literature on the Lysenko controversy provides a solid sense of how politics factored into the latter. Details of the 1948 Presidential election in the U.S. give a good sense of the political aspects of the former, i.e., the milieu in which Zirkle operated. It is ironic that Henry Wallace, among the political figures who suffered most from the anti-communist climate, also became involved in the Lysenko debate.

Wallace represented the Progressive Party against the Democratic candidate Harry Truman, and the Republican Thomas Dewey. Wallace was also the candidate of choice for liberal intellectuals. He had served as Roosevelt's second vice president and then Secretary of Commerce. Wallace was also a scientist, responsible for developing the first commercially viable strain of hybrid corn, which he promoted through the Hi-Bred Corn Company. Hybrid corn transformed agriculture in the Midwest, and the glowing fields of the "corn belt" in the United States would later capture the imagination of Nikita Khrushchev, with important consequences for Lysenko's career.

In the wake of race riots during the Second World War Wallace referred to the irony of fighting fascism abroad while maintaining segregation at home. He believed

---

[71] Ibid, p. 245.

[72] Ibid, p. 167.

[73] See below.

that if you define a fascist as someone who puts money and power ahead of human beings then there were several million fascists in the United States.

> It may be encountered on Wall Street, Main Street or Tobacco Road. Some even suspect that they can detect incipient traces of it along the Potomac.[74]

Toward the end of the war, Wallace visited the Soviet Union and came back with a positive report. He believed Stalin's excesses in the 1930s were growing pains, similar to slavery or the genocide of Native Americans in the U.S. Wallace also gave rosy descriptions of gold mines he did not realize were worked by slave laborers.

As Secretary of Commerce under Truman, Wallace was in touch with Dunn on the translation of *Heredity and Its Variability*. Wallace had been given a book with some of Lysenko's writings while traveling in Central Asia. He agreed with Dunn that Lysenko, "probably isn't a scientist at all, but he has a fresh point of view which may conceivably have in it a small residue of value. At any rate I don't think we should dismiss him completely, even though his temperament is completely unscientific."[75]

By the aftermath of the VASKhNIL conference Wallace had grown more skeptical. When a reporter from the *Washington Post* asked Wallace what he thought of Lysenko's views, he praised Michurin as "the Soviet Burbank." But Wallace also said that once scientists understood the truth about the role of heredity in evolution he doubted it would be what Lysenko thought. Nevertheless, Wallace did believe that Western biologists were so sure Lysenko was wrong that (echoing the opinion of Hudson and Richens) they had not even bothered to try to replicate his work. Wallace said,

> I hope the day will come when in true scientific spirit the results of experiments conducted in this field will be freely exchanged between many different nations.[76]

Shortly after, *New Times*, an internationally circulated leftist journal, published criticism of Wallace's naïve ideas concerning the possibility of reforming capitalism. This was interpreted by the mainstream press as a possible rebuke to his comments on Lysenko.[77]

Wallace became increasingly alarmed by rising Cold War tensions and believed the United States should offer the Russians information on atomic energy in the interest of peace. His discussions with Russian scientists had convinced Wallace that they would soon have it anyway. In his Presidential campaign Wallace proposed a $54 billion "abundance budget," predicated on an end to the necessity for manufacturing atomic weapons.[78] Unsurprisingly, Wallace was soon accused of being "soft on communism." On August 9, 1949 the cover of *Time* magazine

---

[74] "Wallace Defines American Fascism," *New York Times*, April 9, 1944.

[75] Correspondence, Henry Wallace to L.C. Dunn, August 21, 1945. B: D917 L.C. Dunn Papers. Lysenko Controversy in the U.S. #2. The American Philosophical Society.

[76] "Has Democratic Hope 'Genius Will Blossom'," *Washington Post*, September 5, 1948.

[77] "Political Football." *Washington Post*, September 14, 1948.

[78] "Wallace Outlines 'Abundance' Setup," *New York Times*, February 15, 1949.

pictured him as a pied piper leading the gray masses. The caption read: "Henry Wallace: A liberal—or a lollipop?"[79] The school board in Rochester, New York confiscated a textbook, *Twenty Famous Americans*, because it had a chapter on Wallace, and a professor at Evansville College in Indiana was dismissed for supporting his candidacy. Many other academics suffered the same fate once Wallace lost the election.[80]

Perhaps it is no surprise that Dunn was a Wallace supporter. Dunn's name appeared on a full-page advertisement in *The New York Times*, endorsing the Wallace candidacy and calling for a "Century of Peace which the Century of the Common Man demands."[81] Another open letter Dunn signed blamed "anti-Communist hysteria" for the failure to seat a City Councilman from Brooklyn who was a declared member of the Communist Party.[82] Dunn also became embroiled in a controversy concerning the American-Soviet Science Society that had been developing for several years. Dunn had resigned as chairman in 1946, and left the next year for a six-month sabbatical in Scandinavia. Shortly thereafter the Society decided to end its affiliation with the National Council of American-Soviet Friendship because, according to a story later published in the *New York Times*, "Many members looked askance at the Madison Square mass meetings and other activities of the council, which they regarded as political and flavored with propaganda."[83]

The Society was in fact wary of the Council's increased radicalism, but Dunn remained sympathetic to the organization and sought to maintain good relations.[84] In the meanwhile, funding from the Council was replaced by a grant from the Rockefeller Foundation, to be awarded once the U.S. Treasury Department confirmed that the Society deserved tax-exempt status. The Treasury Department had finally responded to the American-Soviet Science Society's application in April 1947, by voicing suspicion that the society was a communist-front organization.[85]

Despite having resigned as chair Dunn became deeply involved in the dispute. As the controversy grew he wrote a letter to the *New York Times* defending one member, the physicist Edward Condon, who was suspected of espionage. He then wrote a follow-up letter defending himself against a reader who questioned why the American-Soviet Science Society and the National Council for American-Soviet Friendship had the same phone number if they were now supposedly separate

---

[79] *Time*, August 9, 1949.

[80] "Charges of Freedom Curbs Rising in Nation's Colleges," *New York Times*, May 29, 1949.

[81] "Display Ad," *New York Times*, October 20, 1948.

[82] "Display Ad," *New York Times*, February 19, 1948.

[83] "Rockefeller Gift Helps Scientists Open Door to Russian Knowledge," *New York Times*, February 8, 1947.

[84] Gormley, *Geneticist L.C. Dunn: Politics, Activism and Community*, pp. 365, 366.

[85] Ibid, pp. 368–9.

organizations.[86] Dunn's defense of Condon and the Society provoked personal attacks in other publications, and finally, in September, 1948, he and his colleagues announced that the Society was shutting down.[87]

## 4.6 The Differences Between Dobzhansky and Dunn; Haldane on the BBC

As all of this was going on, the U.S. government continued to consider Dunn as a candidate for the position of scientific attaché at the American Embassy in London. Over the course of numerous phone calls it became clear that there were deep concerns at the State Department over his interest in Russian biology. Dunn responded to questions having mostly to do with his work with the American-Soviet Science Society. Dunn was obliging when asked to provide further information. In his oral history, conducted a decade and half later, Dunn said he responded to the State Department by saying—"Well, if you haven't got the proper documents in your dossier, I'll send them to you"—and then forwarded them reports and bulletins from various meetings.

The State Department then asked Dunn if, while based in London, he would be willing to make occasional visits to the European continent. Dunn replied: "Yes, I would indeed, and one of the places I think probably should be to the Soviet Union." Dunn was told, however, that the State Department already had someone stationed there. Finally in 1948 Dunn received a letter informing him that, for budgetary reasons, he would not be going. Six months later a professor from the University of Chicago was sent, and Dunn drew the obvious conclusion.[88]

While Dunn's political interests began to hobble his career, Dobzhansky continued to focus on research. He traveled continuously, exploring Peru, Argentina, Ecuador and the rainforests of Guyana and Colombia. The University of São Paulo in Brazil came to feel like a second home.[89] Travel became a gratifying part of Dobzhansky's life as a permanent nomad.

---

[86] "Letters to the Times," *New York Times*, March 18, 1948; "Letters to the Times," *New York Times*, March 19, 1948; "Letters to the Times," *New York Times*, March 6, 1948.

[87] Dunn's efforts on behalf of the American-Soviet Science Society are described in detail in Gormley, *Geneticist L.C. Dunn: Politics, Activism and Community*, pp. 358–83.

[88] The Reminiscences of L.C. Dunn, pp. 794–7; This incident is also recounted in Gormley, *Geneticist L.C. Dunn*, p. 386. Gormley writes that after finding out someone had been appointed in his place, "Dunn said that he drew the obvious conclusion." It is likely that the "too enthusiastic" report Dunn submitted to the Rockefeller Foundation after his 1927 visit also played a role in his being rejected. In *American Hegemony and the Postwar Reconstruction of Europe*, John Krige writes how the officers of the Rockefeller Foundation used the organization's resources and influence to help contain communism at the start of the Cold War. See John Krige, *American Hegemony and the Postwar Reconstruction of Europe* (Cambridge, MA: MIT Press, 2006), pp. 115–51.

[89] Dobzhansky's daughter wrote that after Dobzhansky's initial trip to Brazil in 1943 as part of Roosevelt's Good Neighbor Policy: "The Brazil connection continued for years and gave my father the chance to indulge in his love for the tropics." Coe, in *The Evolution of Theodosius Dobzhansky*, ed. Adams, 24.

Through their correspondence Dobzhansky picked up on the fact that Dunn's Soviet sympathies were interfering with his work. Dobzhansky was troubled that Dunn's defensiveness against anti-red hysteria—as evidenced by his published contentions that Nazi beliefs were worse, and Lysenko's notions never lacked friends among those who found determinism philosophically repugnant—was leading him to actually empathize with Lysenko. In a letter to Dunn from Brazil, Dobzhansky said that Lysenko was a "contemptible cheat" who had obtained backing for "prescientific and at best nineteenth century ideas." "All else is materials for dissertations of future historians," he wrote.[90] Lysenko was guilty; it must be said despite the fact that it helped the arguments of those they disliked. Anything else is like trying to convince oneself that the "snow in New York falls black and turns white in a few days."

> Let us say the truth, and let the chips fall where they may. Maybe the social function of scientists is just exactly that people may look to them for truth, even if telling truth seems at times inexpedient.[91]

But despite his anger, Dobzhansky was unwilling to let their differing points of view on the Lysenko affair destroy their friendship. He concluded:

> Well, enough of this outburst—I hope that we agree on the whole, although I know you are averse from using violent language which I cannot at times avoid. And this is one of cases where I do not even want to avoid it. Violent language has a biological function to serve, and I know of nothing where this function needs more to be served than in the Lysenko case.[92]

If Dobzhansky was frustrated by Dunn, J.B.S. Haldane's reaction to the VASKhNIL conference baffled him. Haldane was immediately besieged with requests to respond, but bided for time. Haldane's usual manner was beguiling charm, as when he arrived to speak at Princeton shortly after the war and said that he would judge American capitalism by its ability to provide him 18½-inch collared shirts.[93] Now, however, he was at a loss for words.

On November 30, 1948, Haldane agreed to participate in a radio program broadcast on the BBC to discuss the Lysenko controversy, and abused his talent for explaining science to confuse his audience. The broadcast was advertised as a "debate," although all four participants were actually recorded separately ahead of time. This was a precaution, according to one commentator, intended to prevent "possible murder" should they meet on the stairs on their way into the studio.[94]

Haldane began by saying that the discussion was "odd," in part because as yet very little was known about the matter: "We are like the jury in *Alice in Wonderland*,

---

[90] Correspondence, Th. Dobzhansky to L.C. Dunn, December 12, 1948. B: D917 L.C. Dunn Papers. Dobzhansky, Theodosius 1948–1949. The American Philosophical Society.

[91] Ibid.

[92] Ibid.

[93] "Haldane in Search of an 18½ Inch Collar," *New York Times*, 26, 1946.

[94] John Langdon-Davies, *Russia Puts the Clock Back* (London: Victor Gollancz Ltd., 1949), p. 78.

considering our verdict before we have heard the evidence."[95] Haldane said that he preferred to wait until a full translation of the VASKhNIL conference was available in English.[96] In fact, as he knew, it already was.

Referring to an article Dobzhansky had written memorializing Vavilov in *The Journal of Heredity*, Haldane distorted its content to imply Vavilov had died a natural death. *The Journal of Heredity* was published in the United States, and the BBC was broadcast in the United Kingdom. Those who were listening could not know better, and those who knew better could not listen. Haldane defended Lysenko with empty non-sequiturs. He said he found it hard to believe the Soviet government would back Lysenko if Lysenko were wrong.[97]

Haldane was well known for his efforts and ability to make science understood.[98] But now people grew suspicious that Haldane's desire to bring scientific ideas to a wider audience was not due to a belief in democracy, but a desire to advance communism.[99] Other geneticists began to complain when he cited their work—with slight alterations non-scientists would not notice—to give credibility to Lysenko's claims.[100] Haldane found a way to twist Lysenko's notions into reason, but once he made them that way they were no longer Lysenko's. At one point during the BBC debate Haldane referred to the possibility of using colchicine to cause mutations in plants to buttress Lysenko's belief in our ability to determine the course of evolution. Unfortunately Lysenko had long condemned the exact treatment Haldane referred to as "torture" and "mutilation." The more Lysenko's work was translated into English, the harder Haldane's task became.[101]

*Time* magazine called Haldane "one of the biggest scientific fish in Communism's net, outside Russia." His colleagues were said to be "watching him closely to see if he would cling to the party line, recently clamped around some very dubious genetics." Science and communism both require religious fervor, *Time* wrote, and the Lysenko situation put geneticists in mind of Galileo. Clearly Haldane would have to choose one or the other.[102]

In the fall of 1948 Haldane published an article in the *Modern Quarterly*, "Biology and Marxism," in which he outlined how Marxist philosophy had contributed to his understanding of the natural world.[103] Haldane struck a balance between genetics and Michurinism by referring favorably to Dubinin's work, as well as

[95] Ibid, p. 88.

[96] Ibid.

[97] Paul, "A War on Two Fronts," pp. 13, 14, 19.

[98] Clark, *JBS*, p. 180.

[99] Clark, *JBS*, 200.

[100] Paul, "A War on Two Fronts," p. 19.

[101] Ibid, p. 30.

[102] *Time*, September 27, 1948, pp. 68–9.

[103] J.B.S. Haldane, "Biology and Marxism," *The Modern Quarterly* 3, no. 4 (1949): 2–11.

Lysenko's notion of "evolutionary leaps."[104] He characterized Social Darwinism as misguided, since eugenicists tended to associate "fitness" with activities like making money. Haldane also pointed out that the "net result" of Adolf Hitler's "activities" had been that "some millions of the 'fittest' Germans were killed, while the remainder have to live in a considerably restricted *Lebensraum*."[105] Haldane wrote that Marxists should focus on the social role of biology: "It should, and could, be part of general culture."

> The countryman should understand the phenomena which he sees every day, from the blossoming of trees to the excavation of mole tunnels, and will certainly be a better agriculturalist if he does so. The town-dweller should have a chance of keeping a few living animals and plants, and of studying them scientifically. … Our health is far below what it might be, and will remain so until the average man and woman have learned to look after themselves at least as scientifically as they can look after a bicycle or a sewing machine.[106]

The primary impediments to the development of a proper scientific attitude among the general public, according to Haldane, were the forces of religion and capitalism. Both feared the revolutionary potential of the application of biology to human society which would, he believed, be greater than the role of physics and chemistry in the Industrial Revolution.[107]

Haldane finally dealt directly with the Lysenko affair on the pages of the same publication several months later, in an article titled "In Defense of Genetics."[108] Here he wrote that "ill-informed" criticism of genetics by Lysenko's British supporters had distorted the issue. He dismissed the practice of labeling biologists with terms like "Mendelist-Weismannist-Morganist," by saying he had no problem calling himself a "Darwinist," even though he rejected the notion of a "struggle for existence," or a "Mendelist-Morganist," even though Mendel and Morgan had both used terms which were mechanist and idealist.[109] Haldane also wrote that if genes were unchangeable then he, "as a Marxist, could not believe in them," and that the results of Lysenko's vernalization technique for transforming wheat had been "repeated so often that it would be very rash to reject it."[110]

Despite his endorsement of vernalization Haldane said he was "skeptical" that acquired characteristics could be inherited. He believed that increased agricultural production in the Soviet Union was more likely the result of collective farming practices. The primary problem was that results of Soviet experiments had not been published "in such a form that they can be repeated," and for this reason it was necessary to improve relations with the Soviet Union. Haldane also claimed that the

[104] Ibid, pp. 6–7.

[105] Ibid, p. 9.

[106] Ibid, p. 10.

[107] Ibid, p. 11.

[108] J.B.S. Haldane, "In Defense of Genetics," *The Modern Quarterly* 4, no. 3 (1949): 194–202.

[109] Ibid, p. 98.

[110] Ibid, p. 200.

reason British geneticists were unable to verify Soviet claims, is that they were "handicapped by their divorce from practical agriculture."[111]

***

Review the quote from St. Thomas Aquinas on the inheritance of acquired characteristics. What does he mean? How does Aquinas' idea square with Dunn's belief that this notion of heredity is *inherently* appealing?

What fears was Dobzhansky addressing in his September 9, 1948 letter to Dunn? To what extent did this—the existence of "red-baiters" who likely welcomed what had happened at the VASKhNIL conference—pose a problem for Lysenko's critics? How was the problem different for Dunn than Muller?

In this same letter Dobzhansky tells Dunn that he is being referred to as a "*mrakobes*" (i.e., "obscurantist fanatic," "religious quack") in the Soviet Union. Why do you think terms like "fanatic," "quack" etc. were invoked so regularly in the controversy? Why was everyone choosing these names in particular?

Is it not odd that the first story in the *New York Times* covering what had happened at the VASKhNIL conference did not even specifically mention that genetics had been banned? Why would this detail—though it would be referred to in follow-up stories—be left out? Was it not the most important fact to report? If not, then what was the story actually about?

What about the analogy presented in a follow-up story—it would be just as if science policy in the U.S. depended upon whether Democrats or Republicans were in office? Do you think that comparison sounded accurate to readers? Do you think the reaction to this statement would be different today?

Since the idea of scientific freedom is so central to the Lysenko controversy, to what extent would you say this is something most people want? This is a question which can quickly lead many directions, but in what ways does the ban on genetics in the Soviet Union in 1948, and the reaction in the U.S., inform this discussion?

What about the reader who wrote that it was hard to be a Lamarckist in the United States? Do you think it is possible that the Lysenko controversy impeded the development of theories of heredity—such as C.H. Waddington's epigenetics—that focus on environmental factors in evolution?[112]

Reviewing the press response to the VASKhNIL conference, and thinking ahead to other examples of this you will read in the coming chapters, why was the (*sic*) opposition between science and religion referred to so frequently? Why did so many commentators seem to find the comparison so obvious, or so useful? Was it fair?

The 1948 UNESCO meeting presided over by Julian Huxley was also described above. What was the relationship between communist demonstrations and protests

---

[111] Ibid, p. 201.

[112] Thanks to Nikolai Krementsov for pointing out to me that the first Russian translations of Waddington's works appeared in Lysenko's mouthpiece, *Agrobiology*. See deJong-Lambert and Krementsov, "On Labels and Issues."

against the founding of Israel? How are these issues related to the Lysenko controversy?

What about Muller's articles in the *Saturday Review of Literature*? What weaknesses—of Muller's and geneticists in general—does the reaction he received reveal? Why were readers so hostile to him?

As for Muller's points about scientific freedom in the U.S., do you agree or disagree with his statements about business, funding, and political loyalty? How do his views—or do they—seem different than they were in 1932?

Best-selling author Howard Fast also criticized Muller. Does Fast's point (Why would the Soviets "rape science" and "cut their own throats"?) sound correct? What about the fact that Fast was not—as he conceded—a scientist? Did his opinion deserve as much consideration as Muller's?

What do you think of Hans Nachtsheim's analysis of the different positions of Nazi and Soviet geneticists vis-à-vis their respective governments?

The theories of Olga Lepeshinskaya and Ivan Pavlov played different roles in the Lysenko controversy. What were they? What purpose did they serve? Are there other examples in the history of science you could compare this to?

Why is the relationship between Trofim Lysenko and his brother Pavlo ironic? Is it? Does it matter that they are brothers, and thus biologically related?

Among the interesting points Conway Zirkle made in "Animals Impregnated by Wind" is that scientific ideas are often conceived long before they are "discovered." What issues does this raise in terms of the Lysenko controversy? How do you think Zirkle's interest in the history of science influenced his reaction to Lysenko?

Zirkle was among the most virulent critics of Lysenko. How were his motives different than Dobzhansky, Dunn, Muller, Huxley or Haldane's? Does it matter that he was a botanist rather than a geneticist?

What do you think of Dunn's opinion that one of the difficulties in challenging Lysenko was that it would require proving a negative, i.e., acquired characteristics are not inherited? Do you think Dunn was correct? And what of his other remarks: Placing Lysenko's actions in context with fascist eugenics, or that Soviet science must be judged differently because it was carried out in a Marxist socialist society?

How does Dunn's view of the problem of proving a negative contrast with Dobzhansky's opinion, referred to in the previous chapter, that scientists should not have to be constantly burdened with disproving "dubious" claims? Who do you agree with?

With regard to *Death of a Science in Russia*, what do you think of Zirkle arranging his own translation of the proceedings of the VASKhNIL session? How does this compare with Dunn and Dobzhansky's translation of *Heredity and Its Variability*? How were their projects similar or different?

This chapter also described the developing Red Scare in the U.S. during these years. In what ways did this environment influence the reaction to Lysenko? Is the Lysenko case unique, or are there other examples where historical context can be as easily attached to how a scientific controversy progressed?

Do you agree or disagree with Progressive Party candidate Henry Wallace's belief that scientific ideas must be freely exchanged? In which instances should they not be?

What do you make of Dobzhansky's frustration with Dunn? What did Dobzhansky mean when he said that Dunn's views were like thinking "snow in New York falls black and turns white in a few days?" Who are you more inclined to agree with—Dobzhansky or Dunn?

Haldane and Zirkle both quote *Alice in Wonderland.* Do you think this is just a coincidence, or is there something about Carroll's book that was particularly relevant to the controversy?[113]

What do you think of Haldane's position in the controversy? What motivated him? How much was he influenced by socialism/communism, versus the belief that all scientific views deserved equal consideration? Was he sincere?

What did Haldane mean when he said he would judge the capitalist system in the U.S. by whether or not he would be able to find 18-½ inch collared shirts for sale in American stores? What was his point?

[113] See also, David B. Searls, "From *Jabberwocky* to Genome: Lewis Carroll and Computational Biology," *Journal of Computational Biology* 8, no. 3 (2001): 339–48.

# Chapter 5
# The "Spitzer Affair"

*For example, during the spring term—I beg your pardon, it was the winter—no, I'm wrong, it was the spring—after campaigning for Wallace he embraced another lost cause; but that was after a period of radicalism during which he asked his freshmen to write on the Moscow Trials, Lenin and Trotsky, the Lysenko theory and other controversial subjects I'm sure they knew nothing about. Some of the students who complained about him said he encouraged discussions of Marxism in his classes. Now I would like you to know, Mr. Levin, that I have no objection to an honest discussion of these subjects, though they certainly don't relate to our "Science in Tech" reader, but I'm sure you'll agree Marxism is specialized subject matter that ought to be confined to mature history or political science courses and not be intruded into freshmen composition. To give you some idea how far astray he went, Mr. Gallegher, our book store manager, called me one morning to tell me that Duffy had placed an order for one hundred and twenty-five copies of "The Communist Manifesto" as supplementary reading matter. I can tell you we soon scotched that.*

—Bernard Malamud, *A New Life*[1]

## 5.1 Some Sympathizers

Lysenko had followers outside of East and Central Europe as well. Michurin societies were formed in Belgium, England, Argentina and Japan. The French *Association Française des Amis de Mitchourine*, who worked with farmers in the countryside attempting to replicate the results of Soviet Lysenkoists, was the most active. A researcher at the National Agronomy Research Institute in Versailles

[1] Bernard Malamud, *A New Life* (New York: Avon Books, 1961), pp. 37–8.

W. deJong-Lambert, *The Cold War Politics of Genetic Research: An Introduction to the Lysenko Affair*, Archimedes 24, DOI 10.1007/978-94-007-2840-0_5,

spent two years trying to create new varieties of tomatoes, using the exact methods prescribed by Soviet Lysenkoists. The work ended in fiasco, but the experimenter lamented: "When an army fights we must not weep for its dead or wounded. I live on with the hope that though I have been injured, I will one day belong to the avant-garde."[2]

An entire special issue of the French magazine *Europe* was devoted to events in Soviet biology.[3] One article, "Storm Over Lysenko," by Louis Aragon, was translated and reprinted in the U.S. publication, *Masses and Mainstream*, to address the political fallout from the Lysenko controversy.[4] According to Aragon, the reason scientists in the West disagreed with Lysenko was because Lysenko agreed with his government: "…the mere fact of approval by the Communist Party in the U.S.S.R., by Stalin, and by the collective-farmers should not be a warrant to compare Lysenko to the Inquisition Court which tried Galileo."[5]

Aragon also argued that "bourgeois" scientists could not accept the idea that new plant and animal species were being created by a new type of scientist, "a peasant 'intelligentsia' of millions of men and women."[6] Lysenko was controversial because he challenged the entire practice and structure of Western science.

> This discussion and what it reveals calls into question bourgeois science at its very core: the men who with all their talents and even genius are still scientists of the bourgeoisie; the manner in which these scientists are recruited; the way in which their science is developed; their *a priori* scorn of practical workers; their belief in the development of science without any links with the people; the dogma profoundly rooted in them and the individual character of discoveries. Yet here, in the name of free discussion of ideas, these are the principles we are told we cannot call into question or discuss.[7]

Aragon said that Lysenko addressed the political implications of Mendelian genetics in order to remove politics from science. Lysenko believed the extension of Malthusian population theory to biology—natural selection—shared a "vital space which is not unrelated to the Hitlerite concept of *Lebensraum*."[8] In denying intraspecific struggle Lysenko refuted the Hobbesian principle, *Homo homini lupus* (man is a wolf to man): "Lysenko once and for all rids biology of the rank weeds of sociology and of politics (in the sense in which his critics understand it)."[9]

---

[2] Bikont and Zagórski, "Burzliwe dzieje gruszek na wierzbie," p. 14.

[3] *Europe*, September–October, 1948.

[4] Louis Aragon, "Storm Over Lysenko," trans. Joseph M. Berstein, *Masses and Mainstream* 2, no. 2 (1949): 22–38; Louis Aragon (1897–1982) was a prominent author, poet, editor and supporter of the Communist Party.

[5] Ibid, p. 27.

[6] Ibid, p. 36.

[7] Ibid.

[8] Ibid, p. 34.

[9] Ibid. An account of the controversy the Lysenko affair provoked among members of the Communist Party in France who were also scientists—particularly Frédéric Joliot-Curie—is provided in Krige, *American Hegemony and the Postwar Reconstruction of Science in Europe*, pp. 129–33. See also the first chapter of LeCourt, *Proletarian Science?*.

The next issue of *Masses and Mainstream* carried a follow-up article, "Revolution in Genetics," by Bernard Friedman, to address the scientific aspects of Lysenkoism.[10] Friedman was a teacher of biology at the Jefferson School of Social Science, a Marxist adult education institute in New York City which operated from 1944–1956, and attracted numerous students interested in Marxist philosophy and labor history.[11] The Jefferson School was the flagship institution of a network of schools around the country operated by the U.S. Communist Party. Friedman taught courses with titles such as "Formal Genetics vs. Michurinism" and "Biology and Marxism." Other materials from the archive of the Jefferson School, such as a bound selection of excerpts from the *Situation in Biological Science* titled *Soviet Biology: From the Discussion on the Lysenko Report*, indicate that Lysenkoism had an extensive influence upon the broader curriculum.[12] For example, in a manuscript circulated to Jefferson School students for their comments and criticism titled, *Child Development*, the author wrote:

> An illustration of what pre-school children can do was related by an American visiting schools in the Soviet Union in 1951. At one kindergarten where children had not yet learned how to read and write, they were already performing agricultural experiments testing Lysenko's theories on small plots of land. The concepts dealt with by these children in describing the purpose and methods of their experiments astounded the American visitor.[13]

Another example of Lysenko's influence is a text, *A Critique of the Bourgeois Concepts of Race and Constitution in Classical Medicine*, presented at a conference on "Science and the Scientist in the Fight for Peace and Socialism," held at the Jefferson School, June 23–25, 1950. The authors criticized Dobzhansky's genetic definition of race, as presented in *Heredity Race and Society* (Dunn was not mentioned), while endorsing Lamarckism, and praising the work of Haldane and Lysenko.[14]

---

[10] Bernard Friedman, "Revolution in Genetics," *Masses and Mainstream* 2, no. 3 (1949): 40–8.

[11] For a history of the Jefferson School see Marvin E. Gettleman, "'No Varsity Teams': New York's Jefferson School of Social Science, 1943–1956," *Science and Society* 66, no. 3 Fall (2002): 336–59.

[12] Course catalogs 1949, 1950, 1951, 1952 and 1956. Addenda. Box 3, Folder 13. Tamiment Library and Robert F. Wagner Archives. New York University; Soviet Biology: From the Discussion on the Lysenko Report [49 pp.]. Box 2, Folder 33. Tamiment Library and Robert F. Wagner Archives. New York University.

[13] Karlson, William, Discussion Manuscript on Child Development (book-length ts, unpublished); Notes on a ms on Child Development … by M.O. (critique). Box 1, Folder 46. Tamiment Library and Robert F. Wagner Archives. New York University.

[14] Medical Student Seminar: "A Critique of the Bourgeois Concepts of Race and Constitution in Classical Medicine." Presented at the J.S. Conference on Science and the Scientist in the Fight for Peace and Socialism [81 pp.]. Box 2, Folder 5. Tamiment Library and Robert F. Wagner Archives. New York University.

The editors of *Masses and Mainstream* described Friedman as a "teacher of biology" and the "recipient of a research grant in cytology by the Carnegie Foundation." This credential would soon be disputed by *Journal of Heredity* editor Robert C. Cook, in a special issue devoted to the Lysenko controversy, published the same year. Referring to Friedman's resume Cook wrote,

> No organization titled "The Carnegie Research Foundation" has been located. The Carnegie Institution of Washington, the Carnegie Foundation for the Advancement of Teaching and the Carnegie Corporation of New York have been unable to identify a Bernard Friedman as being among their grantees. Is it possible that, as is a custom with Party members, he did this work under an alias?[15]

Other details of Cook's critique of Lysenko and his "fellow travelers" will be detailed below.

In his article for *Masses and Mainstream* Friedman stated that in fact no contradiction existed between Lysenko's claims and genetics. The disagreement was only between certain geneticists—such as T.H. Morgan and H.J. Muller—whose beliefs had been heavily influenced by August Weissman.[16] Friedman characterized it as a dispute over the influence of the body on the germ plasm, similar to the "principle of the indivisibility of the atom," which had been based on "the failures of physicists to achieve such a division."[17] Friedman would later publish reviews of Conway Zirkle's *Death of a Science in Russia* and its British counterpart, Julian Huxley's *Heredity East and West*, in *Soviet Russia Today*, accusing both authors of being "self-appointed defenders of our culture," who were unwilling to accept the validity of Lysenko's ideas.[18] Friedman's approach to Lysenko's critics was to portray them as elitists who resisted the idea that a peasant could perform scientific work, while presuming that only they had the right to judge what was significant.

## 5.2 Ralph Spitzer

In 1949 the American Association of University Professors was swamped with a record number of cases dealing with jobs lost for expressing political views or associating with the ideologically suspicious.[19] Legislation for dismissing teachers on grounds of disloyalty passed in Massachusetts, Kansas and Pennsylvania, and educators were forbidden to join organizations identified as subversive in Maryland, New Jersey and New York. In Michigan five faculty members, including one who had run as the Socialist Party candidate for Vice President, were fired. A professor and his

---

[15] *Journal of Heredity* 40, no. 7 (1949): 195 ft.

[16] Friedman, "Revolution in Genetics," p. 41.

[17] Ibid, p. 42.

[18] Bernard Friedman, "Heredity—Fact and Fiction," *Soviet Russia Today*, April 1950, pp. 21–2.

[19] "Charges of Freedom Curbs Rising in Nation's Colleges," *New York Times*, May 29, 1949.

wife—the college librarian—were also subsequently dismissed for holding "ultra-liberal views." Two English professors at the University of New Hampshire, associated with the Progressive Party, resigned under pressure. A professor of Russian Studies at Yale referred to himself as a victim of "thought control" after his plans to invite a Russian composer were cancelled by university authorities. A fellowship granted by the Atomic Energy Commission to a physics instructor at the University of North Carolina was taken back after he was discovered to be an avowed communist. And at Oregon State an Associate Professor of Chemistry, Dr. Ralph Spitzer, was dismissed for—"supporting the genetics teachings of Lysenko, the Russian Communist, who advocates the theory that acquired characteristics can be inherited."[20]

In January, 1949, Spitzer wrote a letter to an academic publication, *Chemical and Engineering News*. His letter was in response to an editorial, "State Science," by the magazine's editor, Walter J. Murphy, which referred readers to the articles H.J. Muller had recently published in the *Saturday Review of Literature*.[21] Murphy indicated that Muller spoke with authority not only as a scientist, but also as someone who had worked as a senior geneticist at the Institute of Genetics in Moscow from 1933 to 1937, and was "personally acquainted" with many of those involved in the "recent purge of Russian scientists."[22] Murphy focused particularly on Muller's contention that the issue was neither limited to Eastern Bloc countries, nor to scientists and science. Though, "the fate of the anti-Lysenkos may seem to have little direct bearing on what goes on outside the Iron Curtain," Murphy argued, "both liberalism and conservatism are fields where the enemies of freedom roam seeking to accomplish their ends undetected."[23]

Spitzer was one of three readers who responded to Murphy's editorial. One author, referring to the link between democracy and scientific freedom, thanked Murphy for having "put your finger on this and … pressed down so effectively."[24] Another accused Murphy of having an "anti-Soviet bias" and argued there was nothing unusual about what was taking place in Soviet biology:

> Certain laboratories are being closed or reorganized, some scientists have been demoted, curricula are to be revised and publications made available for the Lysenkoists. Analogous occurrences are not infrequent in the U.S.A.[25]

---

[20] Ibid. See also Charles H. McCormick, *This Nest of Vipers: McCarthyism and Higher Education in the Mundel Affair, 1951–1952* (Urbana, IL: University of Illinois Press, 1989); Stephen J. Whitfield, *The Culture of the Cold War* (Baltimore, MD: Johns Hopkins University Press, 1991); Ellen W. Schrecker, *No Ivory Tower: McCarthyism and the Universities* (New York: Oxford University Press, 1986). The first article on Spitzer in the *New York Times* reporting Spitzer's dismissal was "Oregon Teacher Out as Lysenko Backer," *New York Times*, February 24, 1949.

[21] Walter J. Murphy, "State Science," *Chemical and Engineering News* 26, no. 52 (1948): 3815. See references to Muller's articles above: Muller, "The Destruction of Science in the USSR," and Muller, "Back to Barbarism Scientifically."

[22] Ibid.

[23] Ibid.

[24] "Letters to the Editor," *Chemical and Engineering News* 27, no. 5 (1949): 306.

[25] Ibid.

Spitzer's letter was the longest, and he went the furthest in directly refuting Muller to defend Lysenko on scientific grounds.

> Contrary to Dr. Muller's assertion that, "despite the pretenses of Communist officials and their followers, this matter is not a controversy between scientists or a dispute over the relative merits of two scientific theories. It is a brutal attack on human knowledge," a perusal of Lysenko's report shows that the issue is largely over matters of biological and technological fact and theory. Are vegetative hybrids possible? Mr. Lysenko has samples. Can the heredity of organisms be changed by changing the environment at an appropriate time and in an appropriate way? The Michurinists have changed 28 chromosome spring wheats to 42 chromosome winter wheats by suitable temperature treatment during several generations. Finally, it is asserted that the Lysenko theory and techniques are far more productive of economic results than the classical theory which is also assailed as being "idealist," a term which in the Soviet Union has roughly the connotations of supernaturalist or unscientific.[26]

Spitzer also argued that purges in Soviet "art, music and economics" were equally exaggerated: "The flurry over the recent criticism of Soviet musicians did not recall the fact that Shostakovich, for one was severely criticized at least as early as 1936, since which time he has won numerous prizes and produced much outstanding music."[27] Rather than arguing that Soviet science operated in a way that was "analogous" to the United States, Spitzer characterized it as superior to "our method of allowing boards of directors, Congress, or the military to decide (often on a smaller scale) which branches of science and which projects to encourage."[28]

On February 8 at 11 a.m., shortly after Spitzer's letter appeared, the president of Oregon State University, Dr. August L. Strand summoned Spitzer and his wife, Therese, to his office. Though Spitzer had been promoted to the associate level, he did not yet have tenure. Therese was an education major involved in a campus group called Young Progressive, which supported Henry Wallace for President. Decades later, a story titled "The 'Red Scare' Comes to Corvallis," was published in the Oregon State alumni newsletter. It described Strand as "habitually sober-faced, forthright, and matter-of-fact." The Spitzers were portrayed as "young" and "nervous."[29] Strand told the Spitzers he was canceling the appearance of a union radical Therese had invited to speak on campus and was not renewing Ralph's teaching contract. When the latter asked why, Strand told him that told it was apparent Spitzer had become more interested in "other matters" than chemistry. Shortly after, Strand dismissed another faculty member, an assistant professor of economics, for similarly obscure reasons.[30]

After learning of Strand's decision to dismiss him, Spitzer brought his case to the campus chapter of the American Association of University Professors and the

---

[26] Ibid.

[27] Ibid.

[28] Ibid, p. 307.

[29] "The 'Red Scare' Comes to Corvallis," *Eclips*. The Oregon State Alumni Association. Available Online at http://alumni.oregonstate.edu/eclips/carry/jan11_2002.html. Downloaded, June 15, 2005.

[30] Ibid.

college faculty council. Both refused to help. In desperation he turned to the press. The story soon made its way from the campus newspaper, *The Daily Barometer*, to the front page of the *Oregonian*. When asked to comment, Strand said he did have reasons for dismissing the Spitzers, but did not see the need to make them public. According to accounts later published in the Oregon State alumni newsletter, Strand was stubborn and responded harshly to criticism. When he received letters challenging his decision to release Spitzer, Strand responded with comments such as "it's too bad you never learned to write in grade school."[31] Spitzer, on the other hand, told reporters he had been promised a good recommendation on the condition he "would go quietly and not raise any fuss."[32]

Finally, on February 23, Strand called a special meeting of the faculty to hear his charges against Spitzer. Strand accused Spitzer of promoting "radical ideas" on campus and teaching Lysenko's theories in his chemistry classes. He then read aloud Spitzer's letter to *Chemical and Engineering News*, and asked,

> Why should a chemist bother to stir up controversy in the field of genetics? I can tell you. It is because he goes right down the party line without any noticeable deviation and is an active protagonist for it. ... Many men in Soviet Russia besides Vavilov have died in concentration camps, or by other means, because they would not accept the untruths which Ralph Spitzer has chosen to espouse. ... Dialectical materialism! A better name would be dialectical murder.[33]

The next day Spitzer issued a response disputing Strand's claim that he supported Lysenko's theories. "Rather," he said, "I suggested that original documents in this controversy be examined in preference to popular commentary in the *Saturday Review of Literature* by H.J. Muller." Scientists, Spitzer argued, needed to work with "first-hand sources," such as the transcript of the VASKhNIL conference, rather than what Muller said about what had taken place. Scientists should not allow themselves to be influenced by "college presidents, Nobel prize winners, or other spokesmen for the dominant scientific theories of the day." Many "impregnable" theories have not "stood the test of time." "Copernicus," for example, "would not have proposed the theory that the earth moved around the sun had he been bound by the authoritative view, dominant for 1,500 years, that the earth was the center of the universe."[34]

The day of Spitzer's rebuttal, the *New York Times* covered the story in an article entitled, "Oregon Teacher Out as Lysenko Backer." The article explained that Strand had dismissed an "associate chemistry professor" for "supporting the genetics teachings of Prof. Trofim D. Lysenko, the Russian Communist." Strand was quoted describing Lysenko as a "charlatan" whose views were opposed by the world's "leading geneticists": "Any scientist who has such poor power of discrimination so

---

[31] "The Spitzer Affair Part II," *Eclips*. The Oregon State Alumni Association. Available Online at http://alumni.oregonstate.edu/eclips/carry/jan18_2002.html. Downloaded, June 15, 2005.

[32] "The 'Red Scare' Comes to Corvallis,".

[33] "Strand and Spitzer Issue Statements on Spitzer's Dismissal," *Chemical and Engineering News* 27, no. 13 (1949): 908, 909.

[34] "Strand and Spitzer Issue Statements on Spitzer's Dismissal," p. 909.

as to choose to support Lysenko's genetics against all the weight of evidence against it is not much of a scientist or has lost the freedom that an instructor or investigator should possess."[35]

Both Strand's statement on Spitzer and the latter's reply were later reprinted in *Chemical and Engineering News*.[36] The same issue also contained letters backing Spitzer from noted chemist Linus Pauling and Alfred H. Sturtevant, a former colleague of Muller in Morgan's "fly room." Though dismissive of Lysenko, Sturtevant questioned the wisdom of "an American university adopting the very policy, of making academic tenure dependent on conformity that we so strongly object to in Russia."[37] Though no one questioned Muller's greater authority to speak on issues in genetics, Spitzer was able to attract sympathy as a victim of the same limits on scientific freedom that Lysenko's critics charged were responsible for Lysenko's success in the Soviet Union. By not resorting, as Muller had, to language echoing the polemics of anti-communist "red-baiters," Spitzer gained support even from those who might also be disturbed that he actually believed Lysenko's theories were credible.

In the longer term the "Spitzer affair" ended Spitzer's academic career in the United States, and the publicity he attracted generated a thick file at the FBI. The next year he was to present a paper on Isaac Newton in Amsterdam. Not realizing they were under surveillance, the Spitzers stopped off in Prague to attend a meeting sponsored by the International Union of Students, an organization listed as subversive. They were arrested by Dutch police upon arrival in Rotterdam and Ralph Spitzer spent a week in jail before being deported back to the United States, without any explanation.[38] Spitzer briefly resumed his career at the University of Kansas City.[39] In 1953 he was asked appear before a Senate Internal Security Subcommittee where he pled the Fifth Amendment after being asked if he was a communist. In 1954 the Spitzers moved to Canada where Ralph attended medical school and taught at the University of British Columbia, from which he later retired. Three decades later Therese Spitzer published a scholarly book on mental abuse entitled *Psycho Battery*.[40]

---

[35] "Oregon Teacher Out as Lysenko Backer," *New York Times*, February 24, 1949.

[36] "Strand and Spitzer Issue Statements on Spitzer's Dismissal," *Chemical and Engineering News* 27, no. 13 (1949): 906–9.

[37] Ibid, p. 936.

[38] "The Spitzer Affair Part II," *Eclips*. The Oregon State Alumni Association. Available Online at http://alumni.oregonstate.edu/eclips/carry/jan18_2002.html. Downloaded, June 15, 2005; Bernard Malamud's, *A New Life* (New York: Avon Books, 1961) tells the story of an academic who relocates to Oregon during the same time period as the "Spitzer affair." The Lysenko affair is mentioned and the book is said to have been inspired in part by what happened to Ralph Spitzer at Oregon State.

[39] Now the University of Missouri–Kansas City.

[40] Ibid. See Therese Spitzer, *Psycho Battery: A Chronicle of Psychotherapeutic Abuse* (Clifton, NJ: Humana Press, 1980).

## 5.3 Muller vs. Dunn; Muller vs. Shaw

Muller agreed with Spitzer's dismissal. If Spitzer supported Lysenko because Spitzer was a communist then he should be fired.[41] As Muller's own experience had shown, Lysenko himself had no tolerance for intellectual freedom. "These people," Muller said, "have blood on their hands; they stink; and there is no use in letting them get away with their pretense that they are representatives of science and culture."[42]

Muller's attitude towards Spitzer was typical of his broader views on communist sympathy and academic freedom. Did a geography professor have the right to teach that the earth is flat?[43] Muller was asked to support faculty at the University of California who refused to sign a loyalty oath stating that they were not members of the Communist Party or any other subversive organization. Muller commented that forcing people to sign oaths was a stupid way to spot communists, but we take oaths all the time—in church, school, when we want to get married. The more important issue for Muller was that communists—just like Nazis or members of the Ku Klux Klan—should not be allowed to teach.[44] He refused to help them.

Muller was also increasingly suspicious that Lysenkoists were going to infiltrate U.S. genetics. He formed the Committee on Public Education and Scientific Freedom in the Genetics Society of America to conduct a public awareness campaign on Lysenko. Muller viewed it as a struggle for the existence of science itself. He had learned however, that actively attacking Lysenko could mean suffering accusations of using "Lysenko-like tactics." Trying to convince the public that Lysenko did not deserve to be taken seriously as a scientist was seen, at least by the readers of *Saturday Review of Literature*, as little better than what Lysenko himself had done.[45]

At the same time, Muller was pitted against Dunn in the leftist press. In an article published in *The Worker*, "Soviet Science is Changing Heredity," the author focused criticism on Muller as a primary figure in the "cold war against the USSR."[46] Dunn's article in the *Overseas News Agency*, "Scientist Finds Russian Dictum on Genetics Purely Political, But Opposed to Nazi Idea," was cited to accuse Muller of telling a "bare-faced lie."[47]

> The American press is so busy spreading such tortured logic that it overlooked the statement made by Dr. Leslie C. Dunn, geneticist of Columbia University. He wrote for the Overseas News Agency that "some people have been apprehensive that official Communist

[41] Carlson, *Genes, Radiation, and Society*, p. 331.

[42] Ibid, p. 375.

[43] Ibid, p. 374.

[44] Ibid, pp. 331–2.

[45] Carlson, *Genes, Radiation, and Society*, pp. 329–32, 375–6.

[46] Peter Stone, "Soviet Science is Changing Heredity," *The Worker* XIII, no. 51 (1948): 5–6, 12.

[47] Ibid, p. 6.

approval and acceptance of Lysenko's views might lead to a Soviet situation similar to that which obtained in Nazi Germany, in which genetics was used to buttress a particular approach to racial differences … the decision in this case would appear to put the Soviet Union in direct opposition to Nazi theory."[48]

Muller and Dunn's actual disagreements were more complicated. They were positioned against one another on the pages of the *The Worker* in part because Muller had been misquoted in *Time* magazine. The editors wrote Muller had pointed out a "doctrinal time bomb that threatens Lysenko's followers." If the heredity of organisms were only shaped by the environment then colonial people would remain inferior indefinitely: "In short, said Muller, it is almost 'Nazi eugenics'." It was a paraphrase of what Muller had said to Lysenko in 1936, but he had never used the words "Nazi eugenics." When he wrote to the editors they refused to print a correction.[49]

In a letter Muller confessed to Dunn that though he regretted the antagonism provoked by his attacks on Lysenko, it was the right thing to do. Also he was comforted by the fact that anyone who had first-hand knowledge of the Soviet Union let him know they agreed. But Muller was angry with Dunn. How could he claim Lysenkoists were upholding a theory that was more "anti-Nazi" than genetics?

It is of course true that they thought that they were doing so but it should be pointed out again and again, as I have done (starting in 1936), that their theory in fact leads to conclusions much closer to those of the Nazis than does the standpoint of reasonable modern geneticists. It is very important that this be pointed out as the Lysenkoists are especially vulnerable on that point and it is unfortunate that a geneticist of high standing should seem to take the other side.[50]

Time for reconciliation was past—"it has been tried for some thirteen years now—and I think all which is left is to call a spade a spade."

p.p.s Did you see the December 26th issue of "The Daily Worker," with its big Sunday magazine feature article, quoting you to refute me? It also attributed to me a long quotation from a non-existent work of mine.[51]

Dunn responded that he had no hard feelings: "I'm sorry you were put to such trouble by my apparent disagreement with some of your views about the Russian situation but I'm very glad you wrote as you did … you may be quite sure that none of this will disturb our personal relations in the least." He agreed with Muller that Lysenko was utterly wrong, but Dunn wondered what explained this, "puffing up of an erroneous view which is going to be so harmful to the Soviet Union."[52]

---

[48] Ibid.

[49] Correspondence, H.J. Muller to Leslie Clarence Dunn, January 17, 1949. B: D917 L.C. Dunn Papers. Lysenko Controversy in the U.S. #4. The American Philosophical Society.

[50] Ibid.

[51] Ibid.

[52] Correspondence, L.C. Dunn to H.J. Muller, February 17, 1949. B: D917 L.C. Dunn Papers. Lysenko Controversy in the U.S. #1.

> Here are hard-boiled people trying to make a collective system work in an economy that is probably not yet ready for it, moved by practical, political and social and economic motives—deliberately choosing a course that will eventually lead to the ruin of Soviet agriculture. What political or other considerations will compensate for the price they must pay for Lysenko? I don't know the answer to this although I have some guesses; but I think it would be short-sighted to take the easy answer that they are all ignorant and evil men since we know that in certain other respects they have been pretty astute. But the main thing is not lose sight of the existence of this problem; and that is why I didn't think that your view that all that's possible now is an autopsy would lead to any further understanding.[53]

Rather than wasting time arguing if Soviet and Nazi views of biology were alike or different, better to try and understand the "mysterious paradox" of Soviet society.[54]

Dunn's dispute with Muller shows how increasingly difficult he found it to balance his life as a scientist, with his role as an activist. As the political pressure surrounding Dunn grew stronger, the members of his department at Columbia turned against him.[55] He began to consider leaving academia and ceasing research. Dobzhansky wrote a letter to Dunn the same month Spitzer was dismissed from his position at Oregon State, encouraging Dunn that he was a leader among scientists.

> To be sure, your political views make you out of tune with the present trend. But isn't this still more of a reason to go on?[56]

A month later Dunn signed a letter sent to the president of the University of Washington who had dismissed three faculty members for being associated with the Communist Party. The letter stated that the case constituted a "shocking repudiation" of democracy and academic freedom, and that the damage caused by their dismissal was "irreparable."[57] The more time Dunn spent on these types of activities the more antagonism he provoked and the less time he had for his work. His twin identities were becoming mutually exclusive.

Meanwhile, Muller's conflict with George Bernard Shaw was much different than his disagreement with Dunn. Shaw was loved by Lysenko's critics for his ability to "stir up the fog without dispelling it," and thus make matters worse.[58] Unlike Haldane, Shaw was not a scientist—he was an artist who hated fatalism and feared what could happen if scientists were not subject to outside authorities. He published an article, "The Lysenko Muddle," in the *The Labour Monthly* where he wrote Lysenko simply asked the question: "Can the State tolerate a doctrine that makes every citizen the irresponsible agent of inevitable Natural Selection?" To Shaw it

---

[53] Ibid.

[54] Ibid.

[55] The Reminiscences of Theodosius Dobzhansky, p. 467.

[56] Correspondence, Th. Dobzhansky to L.C. Dunn, February 3, 1949. B: D917 L.C. Dunn Papers. Dobzhansky, Theodosius 1948–1949. The American Philosophical Society.

[57] "University Urged to Rescind Ouster," *New York Times*, March 1, 1949.

[58] *Journal of Heredity* 40, no. 7 (1949): 207.

was the same as asking: "May you boil your mother to ascertain at what temperature a mature woman will die?" "*Laissez-faire* is dead," he said.[59]

Shaw and Muller published back-to-back articles in *The Saturday Review of Literature* in April, 1949.[60] As described above, the articles Muller had published on Lysenko five months before had not received the reception he expected. He had been accused of engaging in polemics and not restricting criticism of Lysenko to the content of his theories. Since Shaw was not a scientist, his support for Lysenko was grounded more in sociological issues, such as whether the state had an obligation to monitor scientific research, and philosophical questions, such as the potential consequences of believing in inherent biological superiority. Muller, unfortunately, responded in similar terms, and once again made the mistake of not confining the discussion to scientific grounds.

Shaw began his article by saying that to "anyone who knows the ropes" of the Lysenko controversy "the rumpus is laughable." Shaw was a "vitalist," and claimed that Lysenko was a vitalist too, but had to pretend he was a materialist because Marx called his philosophy "dialectical materialism." This, Shaw lamented, "muddles us ludicrously."[61]

> We have a parallel mix-up at home. In the Church of England no candidate for ordination can be inducted to a living unless when catechized by the Bishop he tells the flat lie, which the Bishop knows to be a lie, that he believes without mental reservations everything in the bible literally. ... Lysenko has to tell the flat lie that he is a materialist, and can make the same excuse for what it is worth...[62]

Muller began his response to Shaw by politely noting,

> A man who has done as much to challenge outgrown traditions and to increase social enlightenment as has Bernard Shaw should have the privilege of making some mistakes, even very fundamental ones, without losing our respect. It is an honor to encounter him as an antagonist. We may be prepared to find him expressing some important truths even when he is in error.[63]

Muller wrote that debate over Lysenko's theories would not be settled by "philosophical argument," but "the public has not the patience to be bothered with the intricacies" of genetics.[64] Arguing with Lysenko, according to Muller, would be as pointless as debating William Jennings Bryan—the lawyer for the prosecution in the Scopes Trial—or Henry Norris—the Texas preacher Muller had referred to in one of his previous articles in the *Saturday Review of Literature*.[65]

---

[59] George Bernard Shaw, "The Lysenko Muddle," *The Labour Monthly* 31, no. 1 (1949): 18–20.

[60] George Bernard Shaw, "Behind the Lysenko Controversy," and H.J. Muller, "It Still Isn't a Science: A Reply to George Bernard Shaw," *The Saturday Review of Literature*, April 16, 1949, pp. 10–1, 11–2, 61.

[61] Shaw, "Behind the Lysenko Controversy," p. 10.

[62] Ibid, p. 11.

[63] Muller, "It Still Isn't a Science," p. 11.

[64] Ibid, p. 12.

[65] Muller, "Back to Barbarism Scientifically," p. 9.

Muller sought to separate out the scientific issues by arguing: "It is self-deception to think that the choosing of the mystical or of the genetic view of biological origins would have a direct bearing on our conclusions as to what kind of social system is the most desirable."[66] Muller called on "humanists" and "modern man" in general to cease their hostility towards scientific understanding, and "give up the escapism of believing that caprice is in itself a virtue."[67] He described the history of human civilization as a process by which the associated rise of democracy, technology and education had been mutually reinforcing, creating the conditions for scientists to "objective[ly]" search for truth. He criticized Shaw for joining in what amounted to the most drastic reversal of this process since the Dark Ages and the Inquisition, and believing that he, "a non-scientist, is competent to condemn the hard-won conclusions of geneticists."[68] Muller pointed out the irony that Shaw, "in his day one of the foremost challengers of arbitrary authority," would now ally himself, in the case of Lysenko, with absolutism. He also warned readers that unless they were vigilant, "the doctrine of the infallible state will eventually engulf our culture also."[69]

A few weeks later the editors of the *Saturday Review of Literature* noted that the debate had provoked "considerable comment" from readers. Most of the commentators were against Muller.[70] They criticized him for writing that the public was not patient enough to "be bothered with" an explanation of genetics. One reader claimed that if scientists were capable of producing atomic bombs and biological weapons then everyone had an interest in their work. Moreover, for Muller to claim that Shaw presumed too much, as a non-scientist, in discussing scientific topics, then what right had Muller, as a non-politician, to question the use of power by government? Did the fact that Lysenko's influence was granted to him by Soviet authorities mean that he was wrong?

> I object when the Russians insist that their scientists support Lysenko; I object when the University of Oregon dismisses a professor for supporting Lysenko. But I am not ready to believe that either oppression affects the validity of the particular scientific theory in question.[71]

Other readers felt the same. Why had Muller, by not discussing genetics, "become the fanatical advocate rather than the objective scientist?" Why not be "dispassionate?"[72] Why did Muller expect readers to accept their own ignorance along with his authority: "Is their curiosity about genetic research never to be even partially satisfied unless the sacred text of the geneticists' actual words is perused reverentially?"[73] Muller would have been better off, they argued, by countering

[66] Muller, "It Still Isn't a Science," p. 12.

[67] Ibid.

[68] Ibid, pp. 12, 61.

[69] Ibid, p. 61.

[70] *The Saturday Review of Literature*, May 7, 1949, p. 26.

[71] Ibid.

[72] Ibid, p. 27.

[73] Ibid, p. 28.

Shaw and Lysenko with facts, rather than "his own dogma."[74] If in fact the *Saturday Review of Literature* was not, according to Muller, the proper place for discussing the complexities of genetics, then why bother using it as a forum to discuss Lysenkoism at all?[75]

Muller was caught in a bind. On the one hand, he wanted to criticize Lysenko in a publication that would reach a wide audience. When doing so he was forced, he felt, to address the issue in terms non-scientists could understand. The reaction of readers against Muller indicates a tremendous public mistrust of scientists, particularly when they tried to limit who had the right to debate scientific issues. The idea that "scientific progress" was a process that only scientists should direct seems to have not been as popular with the populace, as it was with scientists themselves.

## 5.4 Heredity East and West—Science

As someone well known for explaining science to a popular audience, Julian Huxley was better qualified than Muller to take on the task of educating the public about the implications of the Lysenko affair. In 1949 Huxley published a book, *Heredity East and West: Lysenko and World Science*, in which he attempted to frame the controversy surrounding Lysenko's destruction of Soviet genetics in terms of the proper relationship between science and government.[76] By the end of the text it was clear he not only believed that the state should not interfere with scientific research, but that genetics should become the basis for official policy and replace religion as an ideology he called "evolutionary humanism." As with Muller, Huxley's argument for a scientific state was influenced by his frustration with the unwillingness of non-scientists to trust his opinion of Lysenko. Huxley used the Lysenko affair to decry popular ignorance of biology, and assert his claim that in the West genetic principles of heredity should be given even greater authority than Michurinism commanded in the East.

Huxley wrote that at first he had thought there must be something to Lysenko's claims, but quickly realized Lysenko and his followers, "move in a different world of ideas," than regular scientists. The form of genetics they call Michurinism, he said, is mostly "ancient superstitions."[77] Though by now most people in the West were aware that something had happened in Soviet biology, few realized what the problem was all about.

---

[74] Ibid, p. 27.

[75] Ibid, p. 28.

[76] Julian Huxley, *Heredity East and West: Lysenko and World Science* (New York: Henry Schuman Inc., 1949). In England the book was published under the title *Soviet Genetics and World Science: Lysenko and the Meaning of Heredity*, by Chatto and Windus.

[77] Huxley, *Heredity East and West*, p. viii.

> This is not surprising, for the whole controversy has been obscured by a fog of misunderstanding, largely resulting from the emotional smoke screen that seems inevitably to envelop any issue concerning the U.S.S.R. Red-baiters have used it as a convenient new stick to beat the Russians with. Communists talk of the resistance of bourgeois science to new ideas. Upholders of free enterprise say "see what happens to science under planning." Believers in state planning point to the necessity for some generally accepted doctrine, including scientific doctrine, to unify society. Pink sympathizers, while avoiding the main issue, make excuses for the Russians' action, or point to the fact that science in Western countries does not enjoy complete freedom. Libertarians let their indignation get the better of them, and confuse the rightness or wrongness of Lysenko's theories with the rightness or wrongness of the drastic methods used to defeat his opponents. Too often, the upholders of one view are ignorant of the different atmosphere of ideas inhabited by their antagonists, and invective has too often taken the place of argument.[78]

As Huxley pointed out, the Lysenko affair could be used to make a variety of arguments, and it was this utility which made the affair all the more confusing.

Huxley explained the basics of Mendelism versus Lamarkcism by pointing out that though constant sun-bathing would have no impact on the skin color of a female's offspring, "Negroes" in Africa developed dark skin due to the selective advantage provided by darker pigment in tropical sunlight.[79] Huxley also attributed the popularity of Michurinism to the distaste of Soviet "political and ideological leaders" for the concept of "innate difference."[80] According to Huxley, Dobzhansky's translation of *Heredity and Its Variability* had "shocked Western scientists." Not only did it show Lysenko's work to be "vague" and "unscientific," but the confidence with which Lysenko justified his views on ideological grounds was taken as an indicator of his power and influence.

Huxley also wrote about Hudson and Richens' *The New Genetics in the Soviet Union*, and said he found their absorption in the relationship between Michurinism and philosophical-scientific developments in the Soviet Union distracting. In other words, the book was not aggressive enough. In Huxley's view the greatest tragedy was that by supporting Lysenko, the Soviet Union rejected the "universal and supranational character of science."[81]

Huxley described most of the arguments surrounding the Lysenko affair in the West as "irrelevant" or "subsidiary." He insisted that issues like the fact that some important geneticists were reactionaries, or Mendel was a Roman Catholic priest whose laws had been abused by Social Darwinists and provided the basis for Nazi eugenics, were red herrings. It confused the issue even more to point out that states everywhere decide how funding for scientific research must be spent, scientists do not always receive the official grants they need, or find it is easy to get their results published. None of this was what mattered. That Huxley also thought "that some

---

[78] Ibid, pp. 1–2.

[79] Ibid, pp. 8–9.

[80] Ibid, pp. 27–9.

[81] Ibid, pp. 34, 35.

geneticists in the U.S.S.R. may have been directly or indirectly 'liquidated'" was also a "subsidiary issue," indicates how strongly he believed that Lysenko threatened scientific freedom: The principle of autonomy meant more to him than the fate of any one scientist.[82] It seems Huxley thought focusing the issue on state infringement of scientific research would be the best of way of uniting scientists in the West against Lysenko.

Huxley also used some of the same arguments against Lysenko and his followers that they enlisted against geneticists. Just as Lysenko's advocates at the VASKhNIL conference had portrayed genetics as un-enlightened, Huxley argued that Lysenko's ascent indicated that, in the Soviet Union, "orthodoxy is once more enthroned." However the Soviets had replaced the "theological orthodoxy" the "Western world emancipated itself" from in the "seventeenth, eighteenth, and nineteenth centuries," with an equally dangerous "social-political orthodoxy."[83] They had, as the another author, John Langdon-Davies, had put it in his own book published that same year, put "the clock back."[84]

Having presented what he labeled "The Ideological Issue," Huxley moved on to "The Scientific Issue." He began with the "question of method," asking: "Do Lysenko and Muller, for instance, both conduct their experiments with adequate scientific precautions, do they publish their data in such a way that they can be checked by repeating the experiments, or that their implications can be fully grasped by other scientists who read about them?"[85] The answer was no, and Huxley believed Lysenko had been able to get away with it due in part to a general level of ignorance among "highly-placed administrators and people eminent in their own walks of life, as well as of the general public," who did not perceive the "distinctive character of science."[86] This problem was not limited to the Soviet Union.

One example of faulty methodology that Huxley cited was forcing self-pollinating plants to cross-pollinate, a practice Lysenko and his followers termed "marriage for love."[87] Rhetoric was another issue. By engaging in polemics and invoking patriotism as a criterion for practice, Lysenkoists "simply do not talk the same language as Western scientists."[88] Huxley went on to discuss other features of Lysenkoism—vernalization, vegetative hybrids, the rejection of statistical methods and Mendel's Laws—he found objectionable.[89] Again, he bristled at the attempt to link Mendelian genetics with Nazi eugenics and racism in the United States: "…Hitler's appeal to Mendelism in support of his racial policy was scientifically quite unjustified …

[82] Ibid, pp. 35–7.

[83] Ibid, pp. 41–2.

[84] Langdon-Davies, *Russia Puts the Clock Back.*

[85] Ibid, p. 64.

[86] Ibid.

[87] Ibid, p. 65 ft.

[88] Ibid, p. 69.

[89] Ibid, pp. 69–88.

most neo-Mendelians are anti-racist … it is untrue that 'the Americans' believe in slavery."[90]

> It is not the Americans as a whole, or the United States as a nation, but only various separate states of the Union that uphold anti-Negro discrimination. Further, the statement that the United States ("the Americans") aim at world hegemony is really not very relevant (even if it were true)…[91]

Even "quite intelligent people" often do not understand, Huxley said, that every theory may in certain details be incorrect. However, even in cases such as the replacement of Newtonian physics with relativity theory, the "whole fabric" of the former did not have to be "scrapped." Though, from Huxley's perspective, it was highly unlikely that Lysenko had anything to contribute to genetics, the greater problem was that the public at large had a limited understanding of how science worked.

To conclude his chapter on the debate's scientific aspects Huxley offered a personal view of Lysenko, based on anecdotes and his own observations. Huxley (like so many before him) compared Lysenko to Savonarola and Rasputin, and said he was "extremely nervous," "unhappy," "unsure," and "shy," yet "forced into the role of leader by a fire within him."[92] He was passionate, and identified criticism of himself as an attack on the Soviet state. Lysenko had a "medieval mind," and was scientifically illiterate.[93] He misunderstood scientific facts and countered them with "bare assertions."[94] Discussing plant physiology and genetics with Lysenko was like talking about calculus with someone "who did not know his 12-times table." Some of his assistants used plant pots without drainage holes, a mistake even an amateur gardener would find amusing.[95] Lysenko genuinely believed that acquired characteristics are inherited. He did not understand genetics—but he did perceive that its "proponents are hindering his work." Rather than admit that "living nature is not so easily taken by storm," however, he preferred to treat genetics "as the enemy, and to root it out from the U.S.S.R., so as to leave the field free for his own ideas."[96]

## 5.5 Heredity East and West—Genetics and Muller

Having concluded his discussion of "The Scientific Issue" by enlisting a biological metaphor to attack Lysenko personally, Huxley next moved on to "Genetics as a Science." Recognizing that his audience probably defined the word "theory" differently than scientists, Huxley began by saying that genetics was—"not 'just

[90] Ibid, p. 87.

[91] Ibid.

[92] Ibid, p. 101.

[93] Ibid, pp. 102–3.

[94] Ibid, p. 103.

[95] Ibid, pp. 103–4.

[96] Ibid, pp. 105–6.

a theory' or a set of untested hypotheses or points of view." He detailed the evidence ("…tens of thousands, of experiments, recorded in scientific journals all over the world … innumerable facts of observation" in numerous fields, "checked, by hundreds or more probably thousands of students…"), and provided examples his readers could relate to. He described chromosomes as playing cards and illustrated inheritance by describing the prevalence of hemophilia among royal families in Europe.[97]

Huxley also addressed some of Lysenko's attacks on genetics, such as that it was based on randomness and chance, and detailed the arguments against Lamarckism. Again, he offered examples intended to appeal to a broad audience. To believe that acquired characteristics are inherited was, Huxley said, "…like supposing that a telegram sent off from Pekin in Chinese will arrive in London already translated into English."[98] He also enlisted a version of the argument Muller had made directly to Lysenko in 1936.

> In parenthesis, it is very fortunate for the human species that acquired characters are *not* readily impressed on the hereditary constitution. For if they were, the conditions of dirt, disease, and malnutrition in which the majority of mankind have lived for thousands of years would have produced a disastrous effect upon the race.[99]

In sum, to reject genetics was, according to Huxley, a "betrayal of the human intellect."[100]

In the next section, "Regimentation of Thought," he described his experience at the World Congress of Intellectuals in Wrocław the previous year.[101] Huxley wrote of a "single system of ideas," in the Soviet Bloc, and portrayed Lysenkoism as part of a trend enveloping every aspect of art, culture, daily life and the natural sciences.[102] Dialectical-materialism constituted a framework as dogmatic as theology, and "Lysenko's success was in large measure due to the fact of Soviet agricultural backwardness."[103] Meanwhile Huxley characterized evidence of "backwardness" in the West as relatively limited.

> In recent times the nearest approach in the Western world to ideological control of science was the legislation prohibiting the teaching of evolution in Tennessee and some other states of the U.S.A. But even this was partial in the sense that it affected only a few states, and only the public institutions in those states.[104]

[97] Ibid, pp. 107–8, 111–3.

[98] Ibid, pp. 134–5.

[99] Ibid, p. 138.

[100] Ibid.

[101] Ibid, pp. 152–3.

[102] Ibid, pp. 155, 156–72.

[103] Ibid, pp. 173, 180.

[104] Ibid, p. 188.

Nevertheless, Huxley did believe that the VASKhNIL conference was "only an extreme and exaggerated manifestation of a general situation"—decreased autonomy for science and scientists.[105]

In the final chapter, "The Situation in Science," Huxley addressed the issue of scientific freedom by arguing for "evolutionary humanism" as a unifying doctrine to guide the Western world. By evolutionary humanism he meant a belief that humans had a duty to bring the "general process of evolution to new heights," a task for which scientists would be primarily responsible.[106] Huxley argued evolutionary humanism should replace religion: "…this evolutionary humanism must be based primarily on science, and it will be the task of the men of science to provide the material basis for the heightened standards of living, and much of the theoretical and philosophic background for the new ideology—what for a religion would be its theological framework."[107]

For Huxley, evolutionary humanism was a way of rolling back the influence of the state upon scientific research, by arguing that the state should become "scientific." While Huxley believed that the government had the right to establish an official science policy, demand that students be educated about science, promote public understanding of science, and combat anti-scientific superstitions—it had no right to determine what was scientifically true or false, or hold science to any standards or orthodoxy outside of the scientific method.[108] Huxley not only decried the fact that Darwinism had not been given its "proper place" in Western education, he argued for a "unified point of view" in how it was understood. Huxley also said he wished the synthesis of genetics and Natural Selection would receive the same level of official support in the U.S. and Great Britain, as Michurinism did in the Soviet Union.[109]

Huxley contended that as societies gained greater knowledge of the "process" of evolution, they could begin to "control certain aspects" of it.[110] In *Heredity East and West* Huxley was not just attempting to rally scientists against Lysenko by characterizing his victory as an attack on scientific freedom. Huxley was also arguing for a state where the work of biologists and an understanding of biological principles would provide the framework for governance. He used the Lysenko controversy to make a claim, on behalf of biologists in the West, for even greater influence than Lysenko commanded in the East. Huxley wanted genetic principles of heredity to become the basis of state policy.

Huxley concluded by attacking George Bernard Shaw ("I heard him once at a luncheon girding at scientists: 'They tell us that the sun is 90 million miles away.

[105] Ibid, p. 195.
[106] Ibid, p. 196.
[107] Ibid, pp. 196–7.
[108] Ibid, pp. 197–8.
[109] Ibid, p. 205, ft.
[110] Ibid, p. 204.

How do they know? They haven't been there'!"), and recounted Shaw's debate with Muller in the *Saturday Review of Literature*.[111] Huxley also took the time to answer the readers who had disagreed with Muller. According to Huxley, it was not hypocritical of Muller, as a scientist, to criticize the political behavior of politicians on a scientific issue. Muller was simply saying that government interference was not compatible with scientific progress. As for the implication that Muller was condescending in not delving into the intricacies of genetics, Huxley argued it would be absurd to think he could do so given the space constraints of a magazine article.[112] Huxley refuted the criticism that Muller had been wrong in not bothering to debate Lysenko, through Shaw, on scientific terms, by pointing out that, "Lysenko…doesn't observe the rules of the scientific game."[113]

According to Huxley, the most egregious criticism of Muller was the contention that he did not understand that his beliefs were "tentative," and might be challenged by future data.[114] This accusation, once again, reflected a larger issue—public misunderstanding of the scientific method. "Too many people," Huxley said, "fail to distinguish between fact, hypothesis, theory, doctrine, and dogma; too many seem to equate pronouncement with proof."[115]

Huxley's tone grew wearier as he catalogued examples of the incapability of the public to separate scientific from philosophical issues when it came to Lysenko. His impatience was evident in statements such as: "Why do people not take the trouble to look up a few facts?"[116] Laymen seemed to believe either that science was as mysterious as magic or that, because theories evolved, there was no such thing as an established truth.[117] Huxley restated one of Muller's points, that it had taken thousands of years for the basis of scientific freedom—democracy, physical techniques, living standards, education, organization—to be constructed.[118] It would take no time at all, however, for it all to be destroyed.

In a postscript, Huxley also responded to Haldane's stance toward Lysenko. He wrote that Haldane avoided the primary issue—"the official banning of Mendelian genetics on the basis of a scientific party line."[119] He also referred to Haldane's article, "In Defense of Genetics," recently published in the *Modern Quarterly*. Huxley praised Haldane for not denying that genes existed, but also raised the question of how Haldane would have fared had he stated this at the VASKhNIL session.

---

[111] Ibid, pp. 210–1.

[112] Ibid, pp. 211–2.

[113] Ibid, p. 213.

[114] Ibid.

[115] Ibid, p. 216.

[116] Ibid, p. 215.

[117] Ibid, p. 216.

[118] Ibid, pp. 219–20.

[119] Ibid, p. 223.

Huxley's criticism of Haldane was more tentative than Muller's, which was probably due to a difference in temperament. Huxley also seemed to appreciate Haldane's subtle criticism of Lysenko when he said that Soviet scientific results had not been published in a "form that they can be repeated."[120] Despite their differences, Haldane and Huxley both agreed that an important issue was the public misperception of what science was all about. For Haldane, this formed the basis of his identity as a Marxist biologist—the belief that individuals should be educated to use biological science to improve their lives.[121] For Huxley the question was not social justice, so much as the hope that if laymen had a better understanding of the scientific method they would accept the premise that science should not be interfered with.

Both Haldane and Huxley had already established successful careers as popular science authors, so it is not surprising that both perceived public understanding of genetics to be a problem the Lysenko affair exacerbated. The difference, however, was that Haldane not only appreciated Lysenko's right to express his views, but also said Lysenko had made him see things he had not recognized before.[122] Huxley, on the other hand, viewed Lysenko as an agent responsible for the subjugation of science by those who were ignorant. In any case, it seems that Huxley interpreted Haldane's "In Defense of Genetics" as a rebuke to Lysenko.[123]

Huxley's condescension towards those who were not scientists was the product of his scientific pedigree, as well as his insecurity. A troubling aspect of this however—also evident in *Heredity East and West*—was Huxley's ideas concerning biological inferiority. While talking about the reasons Africans had darker skin might seem benign, other examples in the book reveal more disturbing details of Huxley's views on race. In another passage he wrote—"For instance, certain differences between a pug-dog and a greyhound, or between a typical Negro and a typical white man, must somehow depend on the influence of the genes concerned on the processes leading to the development of the face and skull."[124] The parallel between "pug-dog" and "greyhound," versus "Negro" and "typical white man" does not seem accidental.[125]

Huxley was also capable of making claims that seem distinctly "unscientific." For example, in an essay in his book *Man in the Modern World*, published the year

[120] Ibid, p. 225; Haldane, "In Defense of Genetics," p. 201.

[121] See references to J.B.S. Haldane, "Biology and Marxism," cited above.

[122] Mikuláš Teich, "Haldane and Lysenko Revisited," p. 560; J.B.S. Haldane, "In Defense of Genetics," p. 195.

[123] In "A War on Two Fronts" Diane Paul writes: "As far as his scientific colleagues were concerned, this article represented his break with Lysenko and the party," p. 34.

[124] Huxley, *Heredity East and West*, p. 4.

[125] For extremely disturbing insight into Huxley's views on race see, Negro Minds. Unpublished MSS. Julian Sorell Huxley – Papers, MS 50, Woodson Research Center, Fondren Library, Rice University.

before *Heredity East and West*, he defended eugenics by saying that societies which ignored eugenic principles suffered. As evidence he cited the rise of Joseph Stalin in the Soviet Union.

> …the theoretical foundations of Communism have prevented the Russians, in spite of their great achievements in pure genetics, from paying proper attention to eugenics. It now appears, however, that they are being confronted with problems, such as the rarity of qualities making for leadership and the inherent difference between a born leader and an ordinary man, which are bound to bring them face to face with eugenics.[126]

Another example was the difference between the Irish who had immigrated to the United States versus those who had remained behind in Ireland. The former, he claimed, had more "initiative" and "adventurousness" than the latter.[127]

If Huxley's motivation for publishing *Heredity East and West* was to explain the Lysenko affair to a general audience, it was also meant to discourage that audience from interfering with science. Huxley believed Lysenko was successful because most people were unable to understand the purpose of science, much less what went on in a laboratory. Lysenkoism was the result of non-scientists taking control of science. The only way Huxley could see of preventing this, was to give scientists greater influence and authority.

## 5.6 The Journal of Heredity

From March 25–27, 1949 the Scientific Conference for World Peace was held at the Waldorf Astoria and Carnegie Hall in New York City. The meeting was sponsored by the National Council of the Arts, Sciences and Professions to promote cooperation and understanding between the United States and the Soviet Union. Attendees from the U.S. included Henry Wallace, Mary McCarthy and W.E.B. Du Bois. Delegates also arrived from Poland, Czechoslovakia, Yugoslavia and the USSR, including Soviet biochemist Alexander Oparin. Oparin was renowned for having developed the "primordial soup" theory to explain the origin of life.[128] Protestors marched outside carrying signs with slogans like "American Not Commie Culture," and two Canadian delegates were seized and deported for "alleged communist activity," while the organization Americans for Intellectual Freedom held a rival conference across town[129]

At the conference, Oparin gave an address in Russian in which he referred to Lysenko's work as an example of the benefits of a "democratic method of research,"

---

[126] Ibid, pp. 41–2.

[127] Ibid, p. 43.

[128] Bowler, *Evolution: The History of an Idea*, pp. 319–20.

[129] "Culture Sessions Center on Conflict of East and West," *New York Times*, March 27, 1949; "Panel Discussions of the Cultural Conference Delegates Cover a Wide Range of Subjects," *New York Times*, March 27, 1949.

and proof that Soviet science "excludes any possibility of its utilization for aggressive purposes." Rather, according to Oparin, Soviet science was "directed to making the life of people better, more joyful, happier."[130] He also spoke of Michurin as one who worked his whole life "toward fulfillment of the dream of mankind: the transformation of our planet into a blossoming, fairylike garden where everything serves the welfare of mankind."[131] Oparin referred to the "disciples of Mendel and Morgan" as those who viewed life as "static and motionless," and were forced to "'await nature's favors,' and with more or less luck grasp 'fortunate chances.'"[132] During the question and answer session which followed, Oparin was asked whether it was true that a number of scientists had lost their jobs for not supporting Lysenko. "No that is not true," he replied, "I can name Lysenko's opponents and tell you where they are working."[133]

The polarity of attitudes evident at the conference—those seeking greater cooperation between the U.S. and the USSR in sharp conflict with those who considered such behavior treasonous—further highlights the dichotomy of responses to Lysenko. The prevailing tension between East and West complicated every position. It is evident that as the controversy continued the rhetoric and strategies of Lysenko and his critics began to appear increasingly similar. Among the best examples of this was an entire issue of *The Journal of Heredity* devoted to the controversy.[134] Muller was on the editorial board, along with Milislav Demerec, a geneticist whose awareness of Lysenko's detrimental impact on Soviet genetics dated back to1936, when the international genetics congress in Moscow was cancelled. The issue, titled "Lysenko's Wonderful Genetics: History and Orientation," consisted of a single article written by the editor, Robert C. Cook, followed by a bibliography on "The Genetics Controversy in the U.S.S.R." Cook's article was headed, "Lysenko's Marxist Genetics: Science or Religion?"[135]

The article began by quoting Vavilov's positive portrayal of Lysenko at the 1932 congress in Ithaca. Cook wrote that in describing Lysenko as an "angry species" then saying that, historically, all progress has been made by "angry men," Vavilov had "underestimated the scope and range of Lysenko's 'anger,' and … overestimated it as a constructive force." According to Cook, Lysenko's "Marxist-Michurinist genetics" was either "the latest thing in science, or the oldest, depending on how we look at it," and the "only scientific discipline in existence today whose validity

---

[130] Alexander I. Oparin, "Science and the Struggle for Peace," *Soviet Russia Today*, May 1949, p. 11.

[131] Ibid, p. 12.

[132] Ibid.

[133] "Panel Discussions of the Cultural Conference Delegates Cover a Wide Range of Subjects." Oparin's speech was later printed in the issue of *Soviet Russia Today*, cited above (pp. 10–2, 28), where it was followed by an article describing how, thanks to Lysenko's methods, citrus trees were now being cultivated further north in the Soviet Union. Amy Schechter, "Citrus Moves North," *Soviet Russia Today*, May 1949, pp. 15–7, 28.

[134] *Journal of Heredity* 40, no. 7 1949.

[135] Ibid, pp. 203–8.

depends not on experiment, but on certification as to purity and truth, in content and concept, by government fiat."[136]

Official Soviet photographs of Lysenko were interspersed throughout the article. In a footnote Cook indicated that they had been sold to the journal "on condition caption is not distorted or factually changed." Given these restrictions, Cook chose to reproduce the official captions under the photos, followed by his own commentary in brackets. For example, a photo entitled "Conference on Wheat Improvement" showed Lysenko surrounded by a group of collective farm workers examining stalks of wheat. He wore a cap identical to those worn by the other men in the group, and the only thing that distinguished him was the medal on his chest. The caption indicated that the conference participants were "chairmen of collective farm boards and agronomists of two districts of the Moscow region," who had been called for instruction on methods for cultivating a new variety of branched wheat. In his bracketed commentary Cook explained that the wheat shown was not a new variety at all, but had in fact "long been known."[137]

> A study of the faces of the "Chairmen of Collective Farm Boards" gives a good idea of the sincere but ignorant peasant types among whom Lysenko has his most enthusiastic support. Here, safe from questions as to the validity of his methods, Savonarola smiles.[138]

Another photograph, entitled "Lysenko's Chickens," showed one of his assistants, Genrietta Korepanova, feeding chickens. Cook commented,

> The unmaterialized (idealistic?) spirit of Lysenko must hover over these premises, for he is visible only in the official legend. How Krepanova's stance and methods of feeding affect the heredity of the chicks is not explained. The pigmented chick in the lower left-hand corner might represent a capitalist Mendel-Morgan influence through action of the genes which Lysenko has outlawed. It is to be hoped that Michurinist influences on this chick will make possible a melanolytic recantation before the emergence of adult plumage.[139]

Another photograph showed Lysenko crouched in a field with two followers, one of whom had a large white beard and wore wire-rimmed glasses. According to the official caption, Lysenko was "measuring the growth of wheat." Cook's caption read: "It is noteworthy that Lysenko, who has interdicted experimental controls and the use of mathematics in biological research needs only eye-power and general impressions to 'measure the growth of wheat.' … The presence of that capitalist symbol, Santa Claus, in the center of the picture is purely coincidental."[140]

Other details of Cook's anti-Lysenko rhetoric are also telling. In a piece entitled, "Lysenko's Marxist Genetics: Science or Religion?", Cook wrote,

> As far as perhaps 95 percent of the population of the world is concerned, what geneticists think about Lysenko is not crucially important. If enough people can be "sold" on the Gospel of St. Marx as revealed by Apostle Trofim, Friar Bacon's hard discipline of rigidly experimental science may be swallowed up in the dialectics of Marx-Engels-Lenin-Lysenkoism. …

---

[136] Ibid, p. 169.

[137] Ibid, p. 176.

[138] Ibid.

[139] Ibid, p. 173.

[140] Ibid, p. 191.

> There are in the aggregate a very large number of people in the world, even in the United States, who still harbor the illusion that genetics is tainted with racism and somehow represents most of the worst features of Presbyterian predestination.[141]

References to Luther Burbank were equally revealing.[142] Cook claimed Burbank enjoyed the same level of support and authority in the U.S. as Michurin and Lysenko in the Soviet Union.

> In this country his name has become a symbol in the popular mind of the great plant wizard, an estimate which is not shared by competent specialists. His contributions to knowledge of plant breeding and genetics are practically nil, and many of his sweeping claims were manifestly absurd.[143]

As Cook's comments on Burbank indicate, he believed that a subversive cohort of Lysenkoists operated in the United States. Indicators of a "Lysenko Fifth Column"—as Cook put it—included English-language translations of the VASKhNIL conference, and a showing of Dovzhenko's Michurin biopic at the Stanley Theater in Times Square, New York City.[144] Cook also tossed in Howard Fast's response to Muller's articles in the *Saturday Review of Literature*, as well as the letters sent from readers which, Cook claimed, betrayed a "leftish tinge," and were "calculated to give the nonbiologist reader the false impression that 'classical' genetics is somehow on the spot."[145]

Other evidence of a "Lysenko Fifth Column" consisted of Bernard Friedman's articles in *Masses and Mainstream* and *Soviet Russia Today*, Alexander Oparin's appearance at the Cultural and Scientific Conference for World Peace and New York, and Ralph Spitzer's activities at Oregon State.[146] With reference to the latter, Cook stated: "Every scientist should of course be free to express his conclusions regarding scientific or other subjects. But there is the question of how much freedom does a Communist actually have?"[147]

## 5.7 He must have realized how stupid it all was…

The next year Dunn helped plan a symposium to celebrate 50 years since the rediscovery of Mendel's Laws.[148] Dunn believed it was important that the event not be merely anti-Russian, and the organizers wondered if they should take Lysenko seriously by addressing him. Did the value of genetics speak for itself—the fact that

---

[141] Ibid, p. 201.

[142] Ibid, p. 193.

[143] Ibid, p. 178.

[144] Ibid, pp. 193–4.

[145] Ibid, p. 195.

[146] Ibid, pp. 195–9.

[147] Ibid, p. 198.

[148] See also Audra Jayne Wolfe, "The Cold War Context of the Golden Jubilee, Or, Why We Think of Mendel as the Father of Genetics," *Journal of the History of Biology* 45, no. 1 (2012).

the U.S. had literally fed Europe with hybrid corn after World War II? Or should the symposium be used as an occasion to counter anti-genetics propaganda? Lysenko's success showed geneticists had done a poor job of presenting their side so far.[149]

Addressing a general audience would require a radical departure from the usual approach. Articles and papers should be composed differently to appeal to a broad audience. One participant, Isadore Michael Lerner, (who would later write the English translation of Zhores Medvedev's *The Rise and Fall of T.D. Lysenko*), said that if the Mendel symposium was intended "as an antidote or a prophylactic against Lysenkoism, then a very unusual course is called for and we shall be actually hurting the cause of genetics by using up our limited time in talking mainly to each other on matters already familiar to us."[150]

> Ever since the beginning of our science 50 years ago many of us have in fact been trying to get it across to the public, but the extent to which the Lysenko movement is being taken up shows how largely we have failed.[151]

Dunn agreed, and believed it was important to avoid giving the impression that work most deserving of reward is that which leads directly to practical application. As he wrote in a letter to another colleague, L.C. Mengelsdorf: "As soon as one begins to pick the winners the invidious distinctions begin to creep in and trouble starts."[152]

In May, Dunn and Muller were both informed of a very strange meeting that had just taken place in Norway. Professor Vsevolod Stoletov, a member of a Soviet Cultural Delegation, presented an informal round-table discussion at the University of Oslo as part of Norwegian-Soviet Friendship Week. The title of his talk was "Guiding Principals of Soviet Genetics." Stoletov was known to be a virulent Lysenkoist, who had vocally opposed the work of geneticists at the 1948 session. The event was not publicized, and the attendees—mostly members of the Genetics Society—were invited by telephone. Two men accompanied Stoletov and there were no outsiders present. He spoke only Russian and questions were answered through an interpreter.[153]

[149] Correspondence, Paul C. Mangelsdorf to Dr. M.R. Irwin, February 14, 1950, cc: Dunn, Huskins and Lerner. B: L563 Lerner, In RE The Mendel Semi-Centennial Symposium (Columbus, O., 1950): corr. 1950–1952. The American Philosophical Society.

[150] Correspondence, I.M. Lerner to M.R. Irwin, March 31, 1950. B: L563 Lerner, In RE The Mendel Semi-Centennial Symposium (Columbus, O., 1950): corr. 1950–1952. The American Philosophical Society.

[151] Ibid.

[152] Correspondence, L.C. Dunn to L.C. Mengelsdorf, February 17, 1950. B: L563 Lerner, In RE The Mendel Semi-Centennial Symposium (Columbus, O., 1950): corr. 1950–1952. The American Philosophical Society.

[153] Correspondence, Jeanne Coyne Mossige to L.C. Dunn, May 14, 1950. Correspondence, Jeanne Coyne Mossige to Muller dated Oslo, May 15, 1950. B: Z67 Conway Zirkle Papers. Mossige, Jeanne Coyne. The American Philosophical Society.

Stoletov began by saying that the science of biology was only appropriate for the study of plants and animals. Man develops according to social laws, and the attempt to apply Mendelism-Morganism to human beings had lead to Nazism, fascism and the belief that there are superior races of individuals. He went on to describe his own work on "reverse-vernalization"—the transformation of spring wheat back into winter wheat. He was vague on the details and responded to specific questions by simply changing the subject. He said he began his experiment with only one grain, and when asked if it was homozygous he became sarcastic: "What is a homozygote? A fruit fly maybe? Or did you ever hear of a homozygous house?"[154]

Next he talked about vegetative hybrids—the usual things about apples grafted onto pear trees, and the potato-tomato combination Lysenko's followers called a "potatomato." A member of the audience said he would be happy to have some seeds derived from the latter plant, but Stoletov replied it would be difficult to send them. Besides, there were excellent potatoes and tomatoes in Norway—he had eaten them himself—so they could make their own hybrids. Stoletov also explained how Soviet scientists had created melons that could survive the harsh climate of Moscow. For four generations they had grafted them onto squash plants. The melon had to be very young, and the squash root older, so that the melon would be influenced by the squash, rather than vice-versa. When someone asked why this had resulted in a hardier melon, rather than a melon-squash hybrid, Stoletov answered distractedly: "Oh, that is done in an entirely different way."[155]

Someone cited a statement Stoletov had made at the VASKhNIL conference, quoted in Zirkle's *Death of a Science in Russia*. Stoletov had said it was now possible to demonstrate experimentally the inheritance of acquired characteristics—even on *Drosophila*. Could he provide details of what he had meant? No, unfortunately, Stoletov replied, he had forgotten what he was talking about.[156]

As the hours wore on, most participants gave up, seeing that it was hopeless to argue with him. A few stubborn audience members hung on, determined that he not believe they agreed with him. Humans had been shaping the evolution of plants and animals through selection since the dawn of civilization, they pointed out. By choosing the largest strawberry or ear of corn humans determined which ones spread and survived without even realizing it. Selection and genetics did not necessarily have to be evil.[157]

Stoletov responded with a story. Once a Mendel-Morganist had told him that if Frederick the Great had been aware of modern genetics he would have bred a race of giant soldiers. This was the danger they ignored.

Who had told him that, they asked? Again Stoletov was evasive, but they insisted on knowing. Finally he reluctantly replied: "Muller."[158]

---

[154] Ibid.

[155] Ibid.

[156] Ibid.

[157] Ibid.

[158] Ibid.

Suddenly, apropos of nothing, Stoletov asked: 'Can anyone here tell me whether or not Muller still believes that acquired characteristics are not inherited?' That was the gist of the question, although the exact wording confused the audience and the translator. It struck everyone as odd and funny—strikingly naïve. A few people took turns answering, saying that as far as they knew Muller had not changed his beliefs. The question seemed, to those in the room, like a silent prayer: "Dear God let Muller still believe so that I can believe."[159]

When one participant followed up by asking if Stoletov was familiar with the work of Japanese geneticist Hitoshi Kihara, he answered curtly: "Of course, do you think we are utterly isolated and uninformed?" And with that he thanked his audience, they thanked him, everyone smiled and Stoletov walked out between the two Russians who had been there, watching him, the entire time.[160]

The talk took place on a Saturday night. The next day a former student of Dunn's, Jeanne Coyne Mossige, wrote letters to Muller and Dunn, informing them of the meeting. It later occurred to her that Stoletov had never, throughout the whole thing, mentioned Lysenko's name—not even once. When someone asked him a question about Lysenko he just skipped it. She looked at the letters again the next day and added a postscript:

I've read this over and the reading of my own words gives me an entirely different impression than did the personal contact with the man. While he was talking he seemed absolutely earnest and sincere; when I reread all this it strikes me that he must have realized how stupid it all was and that his evasions to direct questions were not because he did not know or did not remember but because he knew the standard answers would not be acceptable to us on any basis and he didn't want to appear too much of a fool. He seemed intelligent, very calm and collected and sure of himself, but I suppose his whole presentation, and especially the title "Guiding Principles of Soviet Genetics" was another way of saying "This is my story and I'm stuck with it."[161]

It was the story he was stuck with, but it would not be for much longer.

***

The quote at the beginning of the chapter comes from a novel by Bernard Malamud, which was partially inspired by what happened to Ralph Spitzer at Oregon State University. The character speaking—Professor Fairchild—has strong views on politics, science and academia. What is his point? Do you agree or disagree?

---

[159] Ibid.

[160] Ibid.

[161] L.C. Dunn first met Jeanne Coyne in the early 1930s when she came to work with him as a student at Columbia. Dunn left for a 15 month sabbatical in Oslo, Norway in 1933 and she soon joined him as an assistant. She married and never returned to the States. The Reminiscences of L.C. Dunn, pp. 460–6.

The French reaction to Lysenko was far different than the U.S. Why do you think this was true? Did it matter that Lamarck was French?

Do you agree or disagree with Louis Aragon's points about "bourgeois science?" How have social class, race, economic inequality etc. influenced the history of science? Leaving aside the question of whether Lysenko was "right" or "wrong," is there something to be said for asking the question posed at the VASKhNIL conference: Whose interests does science "serve?"

Speaking of, was there any merit to Bernard Friedman's claim that some of Lysenko's critics were skeptical of him due to his "peasant" background? Keeping in mind that his humble heritage was fundamental to Lysenko's image might it not have provoked an adverse reaction in the U.S. in inverse proportion to how it was received in the USSR? How might the reactions of Muller and Huxley be similar or different on this issue?

Friedman's profession, as listed in the Jefferson School course catalogs, was "Teacher in the N.Y. City High Schools." As far as we know, he had never received an advanced degree in biology. What do you think attracted him to the Lysenko controversy?

Returning again to the quote from Malamud's book, did Walter J. Murphy's editorial have any place on the pages of *Chemical and Engineering News*? Do you agree or disagree with the part of Spitzer's response where he criticized "our method of allowing boards of directors, Congress, or the military to decide (often on a smaller scale) which branches of science and which projects to encourage?"

As for Spitzer's dismissal, was Strand at fault for not being direct about his reasons for not renewing his contract? Was Strand correct that a chemist should not "bother to stir up controversy in the field of genetics?" What about Spitzer's point about trusting "first-hand" sources such as the proceedings of the VASKhNIL conference, rather than listening to what Muller had to say about what had taken place?

Look back at what Ralph Spitzer wrote about scientific funding in the U.S. in his letter to *Chemical and Engineering News* and compare it to what Muller said about the same topic in the *Saturday Review of Literature*. Are they saying the same thing? If not, how are their opinions different?

What about Linus Pauling and Alfred E. Sturtevant's defense of Spitzer. What do you think were their motives? Does it matter that Sturtevant and Muller were rivals?

It is not a surprise that Muller approved of Spitzer's dismissal, but do you agree with Muller's reasons for thinking it was justified? Keep in mind that by this time Muller's own academic career had been hobbled by his political views. How might his have influenced his reaction to Spitzer's fate?

Muller and Dunn also disagreed about Lysenko. What about Dunn's point that, "I think it would be short-sighted to take the easy answer that they are all ignorant and evil men since we know that in certain other respects they have been pretty astute. ... I didn't think that your view that all that's possible now is an autopsy would lead to any further understanding." Who are you more inclined to agree with—Muller or Dunn? Why?

As for Muller's conflict with George Bernard Shaw, do you agree with Muller that Shaw was not qualified to debate him? If so, why did Muller accept the "debate" in the first place? Was Muller's claim that the state must not interfere with the work of scientists legitimate? Why is it ironic that he made it?

What were the reasons the readers of the *Saturday Review of Literature* disagreed with Muller? What was the problem Muller faced? Did his statements merit the reaction they received?

Huxley's approach was to write a book, *Heredity East and West*. But what, aside from all the reasons he felt Lysenko is wrong, did Huxley also address in the work? Do you agree with Huxley's defense of Muller? What about his wish that genetics received the same level of support as Michurnism?

What do you think of Huxley's rhetorical strategy in his attack on Lysenko (e.g., his points about Newtonian physics, using examples like playing cards to illustrate scientific concepts)? What do you make of his role as intermediary between scientists and the public?

Compare Muller and Huxley's views on the correct relationship between science and the state. How do they define the issue differently? What are their solutions?

Among the more surprising names on the list of Lysenko's supporters is Alexander Oparin. Oparin credited Marxist philosophy with being an important influence on his work, and yet his theories were not received the same as, nor have they suffered a similar fate to, Lysenko's. Is this important to the Lysenko controversy? Why or why not?

*The Journal of Heredity* was the first genetics journal published in the United States. What do you think of the fact that Robert C. Cook devoted an entire issue of the journal to Lysenko? Is this a fair role for scientific publications to play when needed?

Why do you think Cook felt so threatened by Luther Burbank, a plant-breeder whose reputation had declined considerably the late 1940s? What was he afraid of? What made him so uncomfortable?

How did the Lysenko controversy influence the program of the Mendel Semi-Centennial Symposium? Would you describe it as positive or negative? What role does commemoration play in how we think about individuals and events in the history of science?

What about the meeting between Vsevolod Stoletov and the Norwegian geneticists? What do you think was going through his mind? Do you think the assessment—"he must have realized how stupid it all was," but "this is my story and I'm stuck with it"—is correct?

# Chapter 6
# Pigs Wearing Booties Earn Scorn of Red Press

> *... In the long run, lying will not prove to be a good method of animal or plant breeding. It is hard to tell whether Professor Kushner believes what he wrote in his article—it seems on the whole more probable that he does not—but doubtless there are breeders who have been led to believe in all earnestness that they can learn the science of breeding from Michurin and Lysenko. Their influence on the breeding work is bound to be like that of witch doctors on the health of their patients.*[1]

## 6.1 First "Fall" of T.D. Lysenko

On August 27, 1951 a dispatch came over the AP wire entitled, "Pigs Wearing Booties Earn Scorn of Red Press." Soviet scientist L.K. Greben had been criticized by *Izvestia* for trying to improve a breed of hogs by putting boots on their feet. Greben was working with a herd of pigs from the Ukrainian plains. The piglets were developing hereditary lameness, a problem Greben attempted to correct by treating them with penicillin and ostrich grease, and making footpaths for them to walk on. After such methods proved fruitless Greben decided to shoe the hogs in special footwear. *Izvestia* wondered why he did not simply introduce new blood into the herd, and quoted a statement he had made at the VASKhNIL session: "The preparation of zootechnicians in the universities on the basis of formal genetic principles brings confusion to the minds of young, specialists to this very day, and hinders us from including properly in general practice the raising of the productivity of animal breeding."[2]

---

[1] Theodosius Dobzhansky, in a review of Kh. F. Kushner, "Michurinist Methods of Obtaining New Breeds of Animals," *Priroda*, 27–34. See, "Animal Breeding Under Lysenko," *The American Naturalist* 88, no. 840 (1954): 165–7.

[2] "Pigs Wearing Booties Earn Scorn of Red Press," B: Z67 Conway Zirkle Papers "An Appraisal of Science in the U.S.S.R." The American Philosophical Society.

W. deJong-Lambert, *The Cold War Politics of Genetic Research: An Introduction to the Lysenko Affair*, Archimedes 24, DOI 10.1007/978-94-007-2840-0_6,

Rumors had begun to spread that Stalin was dissatisfied with Lysenko. "Apparently Comrade Lysenko is getting a swelled head. We ought to set him right!" Stalin had said to a member of his inner circle. "We ought to make Lysenko enjoy criticism," he had scrawled on a report from an agronomist. Hearsay became reality when two articles critical of Lysenko, written by N.V. Turbin and N.D. Ivanov, appeared in an important biology publication, *Botanical Journal*, late in 1952.[3] The former, head of the Leningrad University department of genetics and plant breeding, had supported Lysenko at VASKhNIL by declaring that: "Under cover of their academic titles, fanatical adherents of Morganism-Mendelism commonly engage in essentially empty blather."[4]

Turbin and Ivanov's articles focused on a central tenet of Lysenko's beliefs concerning the transformation of species. Lysenko argued that evolution took place according to leaps and bounds—sudden change, with no intermediate stages. The principle of quantitative change leading to sudden qualitative change (such as water boiling into steam) was a basic principle of Marxism. In *Land in Bloom*, Safonov had described the transformation of nature which would result:

> And so, we changed the climate. It was necessary also to change the landscape to one that contrasted less with the natural requirements of human eyes. …
>
> The physical-geographical and climatic features that had predominated in the past had been a rather irrational combination of mutually contradictory elements. It cost no little effort to introduce some order into this. …
>
> … We think that the more man frees the creative forces of the soil and compels them to work, the more beautiful will the land become. Beauty is the companion of creative life.
>
> On leaving his white cottage hidden in a beautiful garden, a man will walk with a singing heart past fragrant, pearly fields bordered by eternal woods, and the cry of the swan will gladden his heart. He will turn off the road and find himself in the depths of an orchard, among purple plums, and apple and pear trees weighted with fruit. Standing on the edge of a mirrorlike lake, he will not guess that he is standing on the edge of an old ravine that formerly corroded the land.[5]

The utopian visions spun by Lysenko celebrators such as Safonov now seemed increasingly suspect. More evidence against Lysenko was gathered—the fake photograph, the story of the cuckoo, and Olga Lepeshinskaya. Lepeshinskaya's fatal error was that she did not confine her ideas to the study of plants and animals. She became interested in the problem of longevity and declared that the conditions of capitalism—laboring to exhaustion, being poisoned by pollutants on the job—resulted in shorter life spans for the inhabitants of the West. Because life in the Soviet Union was more happy—opportunities for sports, guaranteed vacations, laughter and gaiety—people lived longer, and she envisioned a time when the age limit would exceed 150. She also began proclaiming the life-prolonging powers of

[3] Medvedev, *The Rise and Fall of T.D. Lysenko*, p. 135; Joravsky, *The Lysenko Affair*, pp. 156, 159; Soyfer, *Lysenko and the Tragedy of Soviet Science*, pp. 226–7.

[4] Soyfer, *Lysenko and the Tragedy of Soviet Science*, p. 185.

[5] Safonov, *Land in Bloom*, pp. 487–9.

soda-water baths. Bathing in a tub of sodium-bicarbonate dissolved in water sped metabolism, melted fat, soothed muscles, healed wounds, removed blood clots and affected the chemistry of urine, she said. Her advice resulted in a run on sodium bicarbonate at stores and pharmacies. The theory was later amended to say soda-water enemas worked just as well to account for the lack of private bathing facilities in the Soviet Union. Such notions had now become embarrassing and quietly disappeared from textbooks.[6]

The cuckoo story was just as notorious. Lysenko announced in several lectures that warblers—a colorful member of the songbird family—gave birth to cuckoos after eating caterpillars. Cracks in the cuckoo theory became apparent when a group of graduate students at Riga University in Latvia sent a dissected cuckoo with ovaries clearly visible in the eggs to Lysenko with a tag reading: "What Morganist-Mendelist-Weissmanist stuck cuckoo eggs in this cuckoo?"[7]

A forest near Riga was also the location of a tree widely-cited by Lysenkoists as proof that one species could be generated from another. The tree appeared to be a pine that had grown a spruce branch, and became the topic of a Ph.D. thesis by one of Lysenko's followers. In fact the pine and spruce had been growing side by side, their branches intermingled, and the spruce was cut down. All the locals knew it, and so did the student—but it was still used as evidence by Lysenko. Even worse was the doctored photograph of a hazel tree growing out of a hornbeam, and the graduate student who had been caught gluing parts of two species together.[8] Other frauds were soon exposed, and the month before Stalin's death the period of classical genetics ended as Watson and Crick published their paper on the double helix structure of DNA.

Stalin died on March 5, 1953. His death was reported in the western press with stories covering the impact of his passing on economics, religion, politics and culture.[9] As for science, one article in the *Los Angeles Times* reported that Russian physicians were planning to revive him with a "life machine" which had been under development since 1939.[10] We will never know how Lysenko's career would have progressed had Stalin lived longer, and it seems clear that at the end of his

---

[6] Medvedev, *The Rise and Fall of T.D. Lysenko*, p. 184; Soyfer, *Lysenko and the Tragedy of Soviet Science*, pp. 223–5; Valery Soyfer, "Stalin and Fighters Against Cellular Theory," *Studies in the History of Biology* 3, no. 2 (2011): 83–96.

[7] Soyfer, *Lysenko and the Tragedy of Soviet Science*, p. 210; Gajewski, "Lysenkoism in Poland," pp. 426–7; Berg, *Acquired Traits*, pp. 154–5.

[8] Soyfer, *Lysenko and the Tragedy of Soviet Science*, pp. 229–30, 342, ft. 37; Gajewski, "Lysenkoism in Poland," pp. 427–8.

[9] See for example, "Pope Prays for People When Told Stalin News," *The New York Times*, March 5, 1953; "Dollar Bonds of Russia, Satellites Get Whirl Here on Stalin's Illness," *The New York Times*, March 5, 1953; "Deaths Shroud Stalin Family Life in Mystery," *Chicago Daily Tribune*, March 5, 1953; "Moscow News Causes Stock Price Slump," *Chicago Daily Tribune*, March 5, 1953; "The Stilled Stalin," *Washington Post*, March 5, 1953.

[10] "Magazine Says Reds Hope to Bring Stalin Back to Life," *Los Angeles Times*, March 5, 1953.

life Stalin had begun to plot against him. In any case, though Lysenko would eventually find support from Khrushchev, he would never achieve a similar level of official credibility.

The decidedly impractical results of Lysenko's initiatives continued to become apparent, and his reputation a source of humor. The trees had paid no attention to his theories, and died off at the cost of hundreds of billions of rubles in the forest belts. Lysenko was still alive but a joke circulated that he had died after falling off one of his strawberries. In Poland biologists snickered that Michurin's greatest achievement was cross-breeding an apple tree and a dog: not only was it capable of watering itself, but it barked if a thief tried to steal its fruit. Another joke:

> Academician Lysenko was performing an experiment on the auditory nerve of the flea. He put the flea on his right hand and bade it jump to his left. It jumped, then jumped back again in response to a second command. Carefully he removed the flea's hind legs.
>
> "Jump right!" he commanded, "Jump left!" But the flea did not budge.
>
> "This proves scientifically," he said, "that a flea loses its sense of hearing when its legs are removed."[11]

By this time Haldane had made his break from the Communist Party, a process typically overwrought with contradictions.[12] He wrote his last article for the *Daily Worker*, "They Want to Sterilize the Poor," in 1950, and a few months later reportedly resigned. Still, in 1953 he added his signature to a message of condolence on Stalin's death.[13]

## 6.2 "Eggheads"

As Lysenko lost authority the decline of Leslie Clarence Dunn's career continued. A few days after Stalin's death the State Department decided not to renew Dunn's passport due to the "direction," "domination" and "control" they believed the Communist Party exercised over him.[14] Dunn, as so often before, responded with a letter.

Dunn wrote to U.S. Secretary of State, John Foster Dulles, that his beliefs were no different from communists, or members of the Supreme Court. To abandon them just because they meant he agreed with a member of the former group was contrary to reason and good sense. He supported international scientific cooperation, but it is impossible to provide materials for just judgment of the political

---

[11] Gajewski, "Lysenkoism in Poland," p. 429; "Newspaper Clippings #2 (Lysenko)," B: L563 Isadore Michael Lerner Papers. Newspaper – Z. The American Philosophical Society.

[12] For other examples of Haldane's contradictory positions see Adams, "Last Judgment," p. 459.

[13] Clark, *JBS*, pp. 209–10.

[14] Correspondence, Ruth Shipley to Leslie Clarence Dunn, April 9, 1953. B: D 917 L.C. Dunn Papers. Oral History Records. The American Philosophical Society.

worth of any citizen by dissection and separate assessment of a few facets of his life. What holds them together is what matters, and Dunn believed men governed themselves by reason.[15]

A review he had published of J.B.S. Haldane's book, *New Paths in Genetics*, was part of the evidence against him in his State Department dossier. "I would call attention to the fact that it made unfavorable references to the author's confusion of Marxism with natural science and to his employment of an analogy between society and a living organism which was condemned as antiquated," Dunn wrote.[16]

As a final defense, Dunn pointed out that the USSR had condemned and banned the genetics textbook he had co-authored and edited. Did this not, Dunn asked, indicate that he did not adhere to the Communist Party line? His only purpose in wishing to go abroad was for study: He requested prompt action.[17]

By this time Muller was teaching at Indiana University at Bloomington, where he had arrived after a brief, unsuccessful stint at Amherst College.[18] Muller worked in the same department as Alfred Kinsey, famous for applying field research techniques to study human sexuality. Muller had stopped replying to letters he received from old friends in the Soviet Union, stating that he feared doing so might jeopardize their safety.[19] He also believed the university should construct bomb shelters on campus; but better dead than red.[20]

The day Stalin died Muller was in a very good mood. His secretary told him two gentlemen from the FBI wished to see him. They apologized for bothering him, but Muller replied: "Nothing can bother me today. This morning I heard Stalin died and nothing can make me unhappy today." The agents looked at each other then served him a subpoena to appear before House Un-American Activities Committee.[21]

Muller traveled to Washington, D.C. about a week later. He assumed they were going to ask about Texas or his time in the USSR during the 1930s, but it turned out the reason they had called him was a mistake. His father-in-law had the same name as an agent of the Communist International, and they had mixed up the two. Once the confusion was settled Muller stayed and chatted.

"Why is it true that in our country," they asked, "that so many intellectuals seem to be attracted to or at least do identify with the communistic philosophy? Why do

[15] B: D 65 Dobzhansky Papers. Dunn, Leslie Clarence #1; The Reminiscences of L.C. Dunn, pp. 1145–64.

[16] Ibid.

[17] Ibid.

[18] Carlson, *Genes, Radiation and Society*, pp. 274–303.

[19] Correspondence, SAC, Indianapolis to Director, FBI, September 8, 1952. Subject: Muller, Hermann J. Freedom of Information/Privacy Acts No. 1024867-000. Federal Bureau of Investigation. U.S Department of Justice. Washington, D.C. 20535.

[20] Elof Axel Carlson. Lecture. New York, NY. June 10, 2005; Muller's FBI files contain numerous documents emphasizing that by the 1950s he believed the Soviet Union was even a greater threat to human civilization than atomic warfare.

[21] Carlson, *Genes, Radiation, and Society*, pp. 372–4.

they do it? What is there about it that attracts a man with a university degree or two or three or four of them?"

"Well, you know they made great claims in the old days," Muller replied, "and they are still making them. Claims, although it is harder for them to prove them nowadays."

"What do you mean by the old days?"

"I mean when the Russian revolution occurred, the world was flooded by Russian propaganda literature that attracted many intellectuals."

"What year would that be?"

"Well, 1917. The first revolution was 1917 and the Communists took control in 1918, and from there they went on for some years and there was a lot of propaganda and I daresay that many of the people in the Communist movement of those days did have those aims and they did not realize how sadly it was going to be turned in the opposite direction. They fooled a lot of people, and that is still going on. It fools a lot of people in spite of the overwhelming evidence to the contrary."

"Don't you think that in 1919, when the Russians who were attracted to the Soviet system by the program were absolutely sincere in trying to bring the people a better way of life?"

"Yes."

The committee asked Muller about what had happened in 1948. "Oh they denounced me because I denounced them," Muller replied. Muller said he wished more people could actually be sent to the Soviet Union for a few years so they would know how it is: "More of our so-called intellectual people realize the danger less than others."

"That is it. Eggheads, I call them." one of the committee members said.

"What do you call them?" Muller asked.

"I call them eggheads."

"Unfortunately, the heads are not so easily broken as eggs."[22]

That summer Dobzhansky attended a conference in Hamburg to denounce curbs on scientific freedom in totalitarian states. Many of the participants agreed that much of the fear over security measures in the United States was exaggerated. A physicist from the University of Chicago's Institute of Nuclear Studies asserted that government funding was not an attempt to direct research toward military value. Dobzhansky gave a presentation on biology in the Soviet Union, outlining how the Bolshevik party and the Soviet state had subjugated science to their ends. If anyone had set out to undermine Soviet agriculture they could not have done a better job than Lysenko and his associates.[23]

Zirkle was also on the side of those who believed that the second Red Scare really was just a scare. At an American Committee for Cultural Freedom Conference

[22] HUAC Executive Session Transcripts—RG 233. Box 2. National Archives, Washington, D.C.

[23] "Scientists Demand Greater Freedom," *New York Times*, July 25, 1953; "Government Rule of Science Argued," *New York Times*, July 26, 1953.

at the Waldorf Astoria in New York, Zirkle stated that McCarthyism was overblown and science was not in danger.[24] He was also vigilant in sniffing out any inklings of resurgent Lamarckist fakery in the West. In the summer of 1954 he published a piece in *Science* decrying a subversive trend he detected—attempts to rehabilitate Lamarckism.

> Very rarely in the history of science have efforts been made to propagandize falsehoods knowingly. It should be very interesting to learn how long the attempts will last, how successful they will be, and how many scientists will be deceived.[25]

That same summer at a conference in New York, Dobzhansky kept up the drumbeat against Lysenko by accusing him of spreading old wives' tales.[26] But by now the news had reached the States that his star really was in decline.

## 6.3 Of Khrushchev and Corn, Huxley in Karachi, and Haldane's Exile in India

The trouble began with a dissertation written by V.S. Dmitriyev, one of Lysenko's students. Dmitriyev was not just any student—he was the head of the agricultural planning section of the State Planning Commission. Dmitriyev's position gave him a decisive role in imposing Lysenko's schemes, and he was accepted as a part-time doctoral candidate at Lysenko's institute. His dissertation, written by the staff, described the transformation of rye into rye brome, a type of weed. For every plant there is a weed, he explained, which it turned into when faced with poor circumstances for survival. He, or rather the staff, had subjected the rye to the worst possible conditions—bad soil, too much moisture, close planting, and sowing late with undersized seeds. They then recorded how the rye had degenerated into weeds.

By the time Dmitriyev's dissertation came up for final approval in February, 1954 he had lost his job, and things were not going well for Lysenko. Lysenko turned up to personally defend his candidate with customary sharpness, accusing critics of being Weissmanists—but this time it did not work. The incident was reported in the West under headlines like, "Lysenko Censure Backed by Pravda," and Lysenko's role in the affair was quoted as being a "mockery of Soviet science."[27] One month later, Columbia hosted an academic conference to discuss whether

---

[24] William L. O'Neill, *A Better World: The Great Schism: Stalinism and the American Intellectuals* (New York: Simon and Schuster, 1982), p. 298.

[25] Conway Zirkle, "Citation of Fraudulent Data," *Science* 120, no. 3109 (1954): 189–90.

[26] "Scientists Demand Greater Freedom," *New York Times*, July 25, 1953; "Government Rule of Science Argued," *New York Times*, July 26, 1953; "Lysenko Scored in U.S.," *New York Times*, March 27, 1954.

[27] "Lysenko Censure Backed by Pravda." *New York Times*, March 27, 1954; J.M., "Letter on Lysenko," *Soviet Studies* 6, no. 1 (1954): 105–6; Soyfer, *Lysenko and the Tragedy of Soviet Science*, pp. 211–2, 231–2.

Soviet ideology continued or departed from trends evident in tsarist Russia. At the event, Dobzhansky declared that Lysenko had "destroyed or stultified a whole generation of plant and animal breeders, and caused another generation to be brought up on old wives' tales instead of on scientific knowledge."[28]

Khruschev, it turned out, was behind Lysenko's denouncement, and he now announced his Virgin Lands program intended to raise productivity by cultivating previously uncultivated regions of the Soviet Union. The next year his dislike for Lysenko went from implicit to explicit when he accused the leadership in agricultural science of blocking the use of hybrid corn.

In 1955 Khrushchev met with Roswell Garst, an Iowa farmer who had been introduced to hybrid corn by Henry Wallace. Garst was one of the largest seed-corn producers in the United States, and Khrushchev was enchanted by his fields. Garst nicknamed himself the "ice breaker," and considered it his personal mission to use agricultural cooperation to end the Cold War. Khrushchev and Garst discussed corn sales, and the Soviet Union paid in gold. Garst followed up with several visits to advise Khrushchev on production, as the latter's fascination with corn became legendary.[29]

Khrushchev's interest in Roswell Garst, and his attempt to make corn a principal crop in the Soviet Union, were reminiscent of Lenin's fascination with Luther Burbank. Even as the Soviet Union proceeded with an alternate version of political and economic modernity, the leaders still looked to the United States for examples.[30] Unfortunately, just as interest in Burbank's breeding practices played a role in promoting Lysenko's career, corn and the Virgin Lands scheme would, disastrously, also provide Lysenko with opportunities.

On January 20, 1954, around the same time as Khrushchev was developing his interest in corn, Julian Huxley publicly debated the chief ghost writer of Dmitriyev's dissertation in Karachi, Pakistan. The hall was filled to capacity, with many students and reporters left standing outside.[31] Huxley attacked his opponent's claim that genes are merely a premise, and responded to criticism that he advocated the creation of human castes. In fact, he said, the evidence his adversary cited was Huxley's response to a point Haldane had made: altruistic impulses could never be developed in man due to natural selection. In order to breed individuals who prefer the self-interest of others, it would be necessary to create neuter castes, whose behavior was

[28] "Lysenko Scored in U.S.," *New York Times*, March 27, 1954.

[29] See Harold Lee, *Roswell Garst: A Biography*. Henry A. Wallace Series on Agricultural History and Rural Studies (Ames, IA: University of Iowa Press, 1984); and Richard Lowitt and Harold Lee, eds., *Letters from an American Farmer—The Eastern European and Russian Correspondence of Roswell Garst* (DeKalb, IL: Northern Illinois Press, 1987).

[30] Another more well-known example of this same phenomenon was Stalin's fascination with skyscrapers in the United States. See Vladmir Paperny, *Architecture in the Age of Stalin: Culture Two*, trans. John Hill and Roann Barris (Cambridge, UK: Cambridge University Press, 2002).

[31] Krishna R. Dronamraju, *If I Am To Be Remembered: The Life and Work of Julian Huxley and Selected Correspondence* (Singapore: World Scientific, 1993), p. 109.

not driven by the desire to reproduce themselves. Clearly, Huxley insisted, he was being theoretical—not proposing a method.[32]

The outcome of the debate, according to Huxley, was that he demolished his opponent—a victory confirmed by coverage which appeared in the Karachi newspapers the next day. However though it is true several local publications did declare Huxley victorious, the Pakistani scientist who moderated the debate, Mian Afzal Husain, wrote one newspaper, *Dawn*, objecting to how they had described the proceedings. "Scientific matters," Husain stated, "are not decided by votes."[33]

That summer a farm delegation from the United States reported most agricultural experiment stations in the Soviet Union were ignoring Lysenko's dogmas. On September 2, 1955 the USSR Supreme Court rehabilitated Vavilov due to lack of *corpus delicti*, and a week later the Academy of Sciences listed him on the roster of deceased members. A supplement to volume 51 of the *Great Soviet Encyclopedia* included an entry on Vavilov, though it did not mention his theories, or his relationship to Lysenko.[34]

If acknowledgement of Vavilov's death was late in coming, Lysenko's obituary was premature. Two months after Khrushchev gave a "secret speech" denouncing the arbitrary liquidation, unpredictable terror, and personality cult of the Stalin-era, the *New York Times* reported: "Lysenko, Stalin's Protege, Out as Soviet's Scientific Chieftain." Dobzhansky was quoted, and Muller's reaction was also solicited. Muller said he had been expecting it for a while, but as for whether it presaged a loosening of Soviet dictatorship in general it was too soon to tell.[35] Back in Moscow, Lysenko was snubbed for the first time in many years from the May Day celebrations at the Bolshoi Theater. He would like to have at least phoned in his holiday greetings to Party leaders, but his personally assigned line (one of only 9,999 in the USSR) had been cut off.[36]

In the fall of 1956, Haldane was awarded the Huxley Memorial Medal of the Royal Anthropological Institute, and on November 1st he was invited to give his Huxley Lecture in the Peer's Dining Room at the House of Lords. However the day before, England and France had begun bombing Egypt in response to Gamal Abdel Nasser's decision to nationalize the Suez Canal. Haldane cancelled his appearance, admonishing: "In view of the House of Lords' heavy vote last night in favour of British aggression, I do not propose even to appear to give countenance to so foul an institution by dining there."[37]

---

[32] Ibid, p. 110.

[33] Ibid, p. 109. See also Box 104, folder 8. Julian Sorell Huxley – Papers, MS 50, Woodson Research Center, Fondren Library, Rice University.

[34] Medvedev, *The Rise and Fall of T.D Lysenko*, p. 72; "'Who's Again Who' Listed in Soviet," *New York Times*, September 14, 1958.

[35] "Lysenko, Stalin's Protege, Out as Soviet's Scientific Chieftain," *New York Times*, April 10, 1956.

[36] Soyfer, *Lysenko and the Tragedy of Soviet Science*, p. 243.

[37] Clark, *JBS*, p. 224.

Haldane always said he had decided to leave Britain due to the Suez crisis, but it was more complicated than that. "One of my reasons for settling in India was to avoid wearing socks," he would later say, "Sixty years in socks is enough." [38] Then there was the matter of his wife's arrest. A few days after Haldane's angry letter to the House of Lords his second wife, Helen Haldane, was arrested for being drunk and disorderly. She had stepped on the tail of a police dog after leaving a pub, and become belligerent. Helen Haldane refused to pay a fine and was sent to prison. She was forced to resign from the university, and Haldane decided to resign as well. He then began following up with contacts he had made about positions in India.[39]

Nevertheless, saying it was about politics gave his decision to live in India a romantic nobility it otherwise would have lacked. As he boarded a plane at London airport, leaving for good, he said: "I want to live in a free country where there are no foreign troops all over the place; yes I do mean the Americans." The same day Reuters carried a few anecdotes, titled "Haldane His Own Rabbit":

> Professor Haldane believes in being his own rabbit for experiments. He says:
>
> "It is difficult to be sure how a rabbit feels at any time. Indeed, many rabbits make no serious attempt to cooperate with scientists."
>
> More than ten years ago discussing the possibility that mankind might become extinct, Professor Haldane said:
>
> "If this happens, I venture to hope that we shall not have destroyed the rat, an animal of considerable enterprise which stands as good a chance as any other of evolving towards intelligence."[40]

Huxley said Haldane's attraction to India had to do with the liberal treatment of science and technology he found there.[41] Haldane was also interested in Hinduism. Helen Haldane maintained that by linking man to the animals through evolution, Darwin had converted Europe to the Hindu faith. J.B.S more or less agreed. Haldane referred to the similarity between evolution and the concept of reincarnation, noting that from the Hindu perspective, Darwin had certain qualities of a saint.[42] Haldane also, ironically, accepted and even advocated the caste system. Thinking ahead to the next 10,000 years he said that humans could be cloned and bred to perform special tasks—including travel into outer space.[43]

---

[38] Clark, *JBS*, p. 232.

[39] Clark, *JBS*, pp. 235–6.

[40] "Haldane, Geneticist, Quits Britain For India Which Has No G.I.'s," *New York Times*, July, 25, 1957.

[41] Julian Huxley, *Memories,* vol. II (New York: Harper and Row, 1973).

[42] Krishna R. Dronamraju, "On Some Aspects of the Life and Work of John Burdon Sanderson Haldane, F.R.S., in India," *Notes and Records of the Royal society of London* 41, no. 2 (1987): 222–3.

[43] J.B.S. Haldane, "Biological Possibilities for the Human Species in the Next Ten Thousand Years," in *CIBA Foundation Symposium,* ed. G.E.W. Wolstenhome (London: J. and A. Churchill, 1963), pp. 337–60.

## 6.4 The Final Fall of T.D. Lysenko

On October 4, 1957 the Soviet Union launched Sputnik, the first satellite ever to orbit the Earth. The United States panicked, and things got worse when the Vanguard rocket exploded only two feet off the launch pad at Cape Canaveral two months later. Around the world Vanguard was quickly dubbed "Flopnik" and "Kaputnik." Meanwhile the Soviets sent a second Sputnik carrying a dog named Laika. Laika's cabin came equipped with a television camera, sensors to measure ambient pressure and temperature, as well as the dog's blood pressure, breath frequency and heartbeat. These instruments enabled ground control to determine four days after the launch that Laika was dead. The cabin had overheated, and Sputnik itself passed two days later once its batteries were exhausted. The satellite re-entered the atmosphere after 5 months and 11 days in orbit and burned apart.[44]

Four months earlier, in "Who's Who, What's What?," a current events quiz that was a regular feature in the *New York Times* during these years, question number 11 read: "Moscow reports that Trofim D. Lysenko has been enlisted in the Soviet campaign to increase the supplies of meat, milk and butter. Who is Lysenko?"[45] The question was a testament to the decline in Lysenko's notoriety in the U.S., and the answer was even more humiliating. It read: "Russian geneticist who was the dominant figure in his field under Stalin, later repudiated."

But Lysenko was still credited with one thing: Lulling the United States into complacency. Sputnik showed Americans had underestimated Soviet science. They assumed what Lysenko had done to genetics was symptomatic of what was happening in every discipline, not realizing that in critical areas the Soviet Union was moving rapidly ahead. It was now said Lysenko had done more damage to American ballistics than to Soviet biology, even as McCarthyism ran roughshod over U.S. academia.[46] But all of that was about to change.

Two weeks after Sputnik's immolation, Khrushchev told a conference of agricultural personnel that few scientists understood the soil like Lysenko. He accused those who opposed him of "sitting with their hands folded like saints," turning their backs on the people.[47] Khrushchev said he ran a competition between Lysenko and another academician, Nikolai V. Tsitsin, in the fields near his dacha. Lysenko applied a special organo-mineral mixture he had invented as fertilizer and, Khrushchev bragged, achieved a greater yield. Good fortune again—just like when his father had buried seeds in sacks under snow three decades before.

[44] Anatoly Zak, "The True Story of Laika the Dog," Available online at http://www.space.com/news/laika_anniversary_991103.html. Downloaded, January 21, 2005.

[45] "Fifteen News Questions," *New York Times*, July 21, 1957.

[46] "Party Dogma Found to Hinder Some Areas of Soviet's Science," *New York Times*, July22, 1959.

[47] "Khrushchev Pays Lysenko Tribute," *New York Times*, April 11, 1957.

Khrushchev was impressed and Lysenko received his seventh Order of Lenin for his sixtieth birthday.[48]

Meanwhile in the West a group of French biologists claimed success using injections to induce inherited characteristics in ducks. Out of 26 ducklings hatched, 70% had acquired beak and foot characteristics their parents had received from a syringe. "If the French research has succeeded in causing transmission of acquired characteristics from one generation to the next," *The New York Times* reported, "it would seem to offer some confirmation of the ideas of the Russian Prof. Trofim D. Lysenko." Lysenko cited the ducks and claimed vindication.[49]

The tenth International Congress of Genetics was held in Montreal in August, 1958. The organizers expected that a large number of Soviet geneticists would attend, however at the last minute several cancelled. The Soviet delegation then sent a new list of papers to be delivered, which by their titles seemed to indicate Lysenko might be back in control. One biologist from the Academy of Sciences in Moscow presented his research using blood transfusions to transfer characteristics from colored to pure-bred white chickens. They claimed the characteristics were inherited.

The delegation was led by Professor Stoletov, who had given the presentation at the University of Oslo in 1950 that Jeanne Coyne Mossige had described to Muller and Dunn. Stoletov now insisted that politics played no role in Montreal—"Professor Trofim Lysenko himself did not know who was to come here"—he declared. At the closing session, a statement was read from the Permanent International Committee for Genetics offering deep sympathy for scientists whose governments may have prevented their attendance.[50]

A session of the Central Committee in Moscow in December, 1958 confirmed the impression given in Montreal. A Party leader from Moldova complained about the criticism of Lysenko that had appeared in *Botanical Journal*. The editors should be replaced, Khrushchev replied—and they were.[51]

But backing from Khrushchev was not the same. His obsession with corn was a source of humor; one joke went that his solution to the Suez crisis was to fill up the canal and plant maize.[52] His promises that socialism would be built in a matter of

[48] Soyfer, *Lysenko and the Tragedy of Soviet Science*, p. 254.

[49] "French Say Chemical Injection May Change Heredity of Ducks," *New York Times*, July 23, 1957; "Lysenko Defends His Theory Anew," *New York Times*, December 15, 1957.

[50] "Russians Revise Genetics Papers," *New York Times*, August 21, 1958; "Blood Types Tied to Some Diseases," *New York Times*, August 22, 1958; "Soviet Criticized on Genetics Issue," *New York Times*, August 28, 1958.

[51] "Party's Meeting in Soviet Lively," *New York Times*, January 11, 1959; Soyfer, *Lysenko and the Tragedy of Soviet Science*, p. 262.

[52] Putrament, "How I Became a Lysenkoist," p. 444.

years fell flat far short of utopia. Khrushchev was embarrassing, and his re-discovery of Lysenko was telling. According to Zhores Medvedev, a chronicler of the "rise and fall of T.D. Lysenko," supporting Lysenko was 10–12% of the reason Khrushchev was removed from power.[53] But that would come later.

Just as Khrushchev was no Stalin, genetics was also not the same science it had been 10 years before. News of DNA and the genetic code were widely published in the Soviet Union. Lysenko was defensive; he said it was just an evasion, and warned against replacing biology with chemistry and physics. But advances made it far harder to remove the language of Western biology from scientific texts. Impassioned calls to overtake the United States in meat, milk and butter by way of Michurinism inevitably involved a stumbling detour into what was being done on the other side of the Iron Curtain. Still, in August, 1961, 13 years after the session in Moscow, Lysenko was reappointed head of the Lenin All-Union Academy of Agricultural Sciences.[54]

## 6.5 Conway Zirkle

As we reach then end of this book it is worth returning to a topic we began with: Conway Zirkle. In 1959 Zirkle published *Evolution, Marxian Biology, and the Social Scene*, which functions as a bookend to the work he had published exactly a decade before, *Death of a Science in Russia*.[55] By comparing these two publications we discover something—at least if you are approaching Zirkle vis-à-vis Lysenko—unexpected. This surprise enables us to begin considering a broader research question that the Lysenko controversy uncovers.

Two of the most significant developments in biology by the late 1950s and early 1960s were a profound shift in how people in the United States thought about race, and the attempt to resurrect eugenics as valid policy that ought to be pursued. The former, thanks to the attendant social transformation enacted by the Civil Rights Movement, occupies a far more prominent position in the history of the U.S. than

[53] Correspondence, Zhores Medvedev to David Joravsky, October 20, 1964. Correspondence, David Joravsky to H.J. Muller, October 27, 1964. B: L563 Lerner. Medvedev, Z.A.—materials 1962–1964. The American Philosophical Society. Medvedev's estimate was probably not meant to be precise, but more a reflection of his own perception of how much it cost Khrushchev to support Lysenko. In his biography of Andrei Sakharov, Richard Lourie writes that item 14 on the 15 counts listed against Khrushchev was, "failing to heed Academician Sakharov's protest against 'Lysenko's nonsense.'" Richard Lourie, *Sakharov: A Biography* (Hanover, NH: Brandeis University Press, 2002), p. 182.

[54] Joravsky, *The Lysenko Affair*, pp. 169–86; Soyfer, *Lysenko and the Tragedy of Soviet Science*, pp. 253–70.

[55] Conway Zirkle, *Evolution, Marxian Biology, and the Social Scene* (Philadelphia, PA: University of Pennsylvania Press, 1959).

the latter. However since this is a book about how biologists thought and behaved, we may treat them as equally significant.

So what does this have to do with Zirkle's book? Let us start with a diagram the author provided in the first chapter.[56]

DIAGRAM OF INTERACTION OF CULTURAL AND BIOLOGICAL EVOLUTION

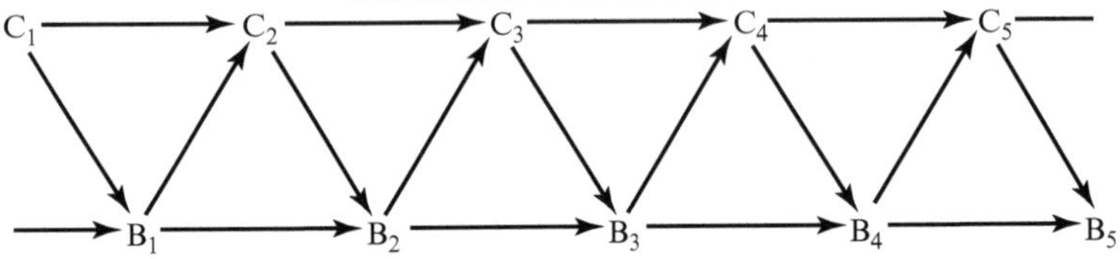


$C_1$, $C_2$, . . . $C_5$ represent successive cultural stages, and the arrows between, them the internal logic of cultural development. $B_1$, $B_2$, . . . $B_5$ represent successive stages in our biological evolution and arrows between, the dependence of each stage upon its predecessor. $C_1 \rightarrow B_1$, $C_2 \rightarrow B_2$, etc. indicates the biological selectvity of human culture. $B_1 \rightarrow C_2$, $B_2 \rightarrow C_3$ etc. represents the fact that culture always passes into the custody of those it selects and that its growth is conditioned by their abilites and activities.

What does that mean? Zirkle's drawing was meant to illustrate the "great man theory" of historical progress. This, of course, is the precise opposite of Marxism. In an evolutionary theory of history premised on how large populations evolve individuals are of no interest. Yet some might believe that certain individuals (such as themselves) deserve credit for civilization.[57]

Could such views ever be reconciled? No. So what Zirkle did was go after any "scientific" proof which he saw being interpreted in a way intended to disprove his favored theory. Though a fair amount of the text in *Evolution, Marxian Biology and the Social Scene* refers to Lysenko, the "death of a science in Russia" had long left center stage as the focus of Zirkle's (apparent) concerns. This book was much more about biological superiority, or better—why some refused to accept it as fact—than it was about what Zirkle had taken to calling "the Lysenkoids."[58]

There are a couple of important clues betraying Zirkle's agenda. The first was his frequent use of the word "equalitarianism."[59] The significance of this term will become more obvious below when we look at a book Dobzhansky was asked to review a couple of years later. Suffice it to say for now that for Zirkle, "equalitarianism"

[56] Ibid, p. 52.

[57] See also Conway Zirkle, "Some Biological Aspects of Individualism," in *Essays on Individuality*, ed. Felix Morley (Philadelphia, PA: University of Pennsylvania Press, 1958), pp. 37–62.

[58] Ibid, p. 123; See also Loren Graham's reference to Zirkle in *Science and Philosophy in the Soviet Union*, p. 473, ft. 3.

[59] Ibid, pp. 106, 424, 443.

was a handy way of implying that anyone who believed in human "equality" was a communist. As usual, the best way to get away with something (e.g., the larger –ism kerfuffle: "Mendelism-Morganism-Weismannism-Lysenkoism…") was to accuse others of it, and analyze their behavior. Zirkle quoted Marx and Engels' use of the word "nigger," even as he used the word "race" in a way that was shockingly indiscriminate.[60] When describing how a totalitarian government might affect the genetic endowment of the population they terrorized, he wrote,

> …the differential butchery of the Chinese by those now in power may well have altered the genic endowment of the Chinese race. Another form of totalitarianism also had its evolutionary effects in the Hitlerian concentration camps, camps that lowered the genic endowment of the human race.

Though one might would not be surprised by say, a journalist like Waldemar Kaempffert or a playwright like George Bernard Shaw being blissfully unaware that it is impossible to define the "Chinese" as a race if you then define "humans" as a race, for a biologist in 1959 to display such ignorance is shocking.[61] Zirkle's elementary error brings us closer to his real motivations.

These are most apparent if we focus on Zirkle's treatment of Darwin and—even more significantly—Lamarck. At this point we must ask: What bugged Zirkle so much about Lamarck? Here was a man who had been dead for over a century, and yet seemed to upset Zirkle even more than his nemesis from a decade before—Lysenko. Or had it always been this way? Indicators that the Lysenko affair had really been, for Zirkle, the "Lamarck affair" all along, are abundant.

Aside from bemoaning the conflation of Lamarckian and Darwinian ideas among communist philosophers, while ignoring the "eclipse" of Darwinism in the United States and Great Britain (a phenomenon in large part attributable to Darwin's ultimate conversion to Lamarckism), Zirkle also derided Lamarck to an extent on-par with his excoriations of Lysenko.[62] Just as in *Death of a Science in Russia* Zirkle had called Lysenko a "simple charlatan" who was not only an "ignoramus," but "stupid" and "bigoted" as well, he now said Lamarck was a "duffer," "ridiculous," "absurd," "embarrassing" and—most damning of all—"not important." Moreover, his "writings on meteorology mean little," "his geology plainly belongs

[60] Ibid, pp. 93, 110.

[61] See Julian Huxley and A.C. Haddon, *We Europeans: A Survey of "Racial" Problems* (New York and London: Harper and Brothers, 1936). Though Huxley was far from enlightened when it came to "race," his book is typically considered as part of a process of replacing the anthropological, with a new biological, view of race. See also Dunn and Dobzhansky, *Heredity Race and Society*, cited above.

[62] The views of Marx, Engels, Prezent, Lysenko et al. were not, of course, nearly as simplistic as Zirkle portrayed. See also Kouprianov, "The 'Soviet Creative Darwinism' (1930s–1950s)"; For a comparative account of Darwin's reputation in Russia, see Yasha Gall and Mikhail B. Konashev, "The Reception of Darwin's Theory of Evolution in Russia: 1920s to 1940s," in *The Reception of Charles Darwin in Europe,* vol. 2, ed. Eve-Marie Engels (London and New York: Continuum International Publishing Group, 2008), pp. 502–21.

to the lunatic fringe," he was "unable to grasp even simple physics," and—deservedly—he "was laughed at."[63]

In the case of Lysenko—if we took Zirkle straight—we would say his barbs were intended to defend the integrity of science against a "pseudoscientific charlatan." But that does not make any sense with regard to Lamarck. Zirkle was a historian of science; he knew what the rules were. There is none of this whiggishness to be found in his *American Naturalist* articles or American Philosophical Society essay on Lamarck and the inheritance of acquired characteristics. But the audience for *Evolution, Marxian Biology, and the Social Scene* was totally different. Zirkle was not writing history, he was making an argument about the present.

And here is what upset him: Zirkle's hobbyhorse was a return to a neo-Malthusian eugenics that would have been met with approving nods from any eminent "eminent Victorian" advocate of Social Darwinism. Zirkle wrote,

> It is easy to make fun of zealots and eugenics had its full fringe of cranks. The early eugenicists, moreover, did not realize all of the difficulties of their program. … A serious eugenic program, however, could avoid all such extravaganzas, but even if it did, it would be faced with practical obstacles. The only possible method of enforcement might well require an educational and ethical level well beyond the reach of the generality of mankind. Any possible eugenics program, in consequence, must be very modest. At first, all we can hope for is to correct the widely propagandized misinformation on the subject.[64]

So the idea was—let us start where we left off before it got so ugly, and let "rational" heads prevail. Most people are not smart enough to understand why this is important, so we must choose cautiously. We will start with basic steps such as anti-propaganda, before moving on to policy. At that point we will select the best and breed them. We will do it right this time.

Did the inheritance of acquired characteristics undermine this? Maybe, but not necessarily. The point is Zirkle thought it did, and that is why he went after Lysenko. We can encounter someone walking through this exact same door from the opposite direction by consulting "Professor Zirkle's Vitriolic Attack on Lamarck," an article published by Harry Gershenowitz in a 1984 issue of the *Indian Journal of History of Science*.[65] The name Lysenko comes up not once in this article. It is a parallel world. Here is the abstract:

> For forty-four years Conway Zirkle (1895–1972), long-time professor of botany at the University of Pennsylvania, spewed envenomed literature of shameful inaccuracies concerning the professional and personal character of the French scientist and evolutionist, Jean Lamarck (1744–1829). Above all, he stressed the purported oddity and absurdity of the characteristics of Lamarck. Professor Zirkle, a member of the neo-Darwinian donnish elite, misinterpreted the evidence relative to Lamarck's scientific contributions to the development of modern evolutionary theory. This writer analyzed six major writings of Professor

[63] Zirkle, *Evolution, Marxian Biology, and the Social Scene*, pp. 72–6.

[64] Zirkle, *Evolution, Marxian biology, and the Social Scene*, pp. 457–8.

[65] Harry Gershenowitz, "Professor Zirkle's Vitriolic Attack on Lamarck," *Indian Journal of History of Science* 9, no. 13 (1984): 261–71.

> Zirkle and discovered that only two sources of Lamarck's voluminous works were used as material to attack Lamarck's conclusion as to the causes for biological evolutionary changes. Professor Zirkle in six writings assaulted Lamarck's respectability, wisdom, experience and judgment by the replete use of two examples extracted from only three pages of Lamarck's many publications.

Zirkle hated Lamarck.

The final pen stroke of this argument appears in a review of a book, *Eugenics and the Progressives*, by Donald K. Pickens, that Zirkle wrote in 1969, three years before he would die. It was published in a special issue of *Annals of the American Academy of Political and Social Science* on the theme of "Protest in the Sixties." The editors had selected essays focusing on the fact that one of the distinct features of the social movements that had emerged during the decade was that they were led by the younger generation. Hippies, liberals, anti-war activists, Black Panthers, feminists, and other members of the New Left subculture; you can guess what Zirkle thought of such people. Zirkle's doom-saying enthusiasm for *Eugenics and the Progressives* highlights how nice it must have felt for him to realize there were others who believed the now highly conservative cause of eugenics had simply been misunderstood, and deserved a second chance.[66]

After noting approvingly how Pickins had remained "free from any of the emotional perturbations that are often found in those who are concerned with the subject," Zirkle gave his own rationale for why it was time to rethink eugenics.

> We now have troubles, and some of the topics that concerned the eugenicists have come again to the fore. In spite of our unprecedented labor shortages, we have growing relief and spreading poverty. Many of our large northern cities have had race riots and the accompanying looting and burning. Crime is steadily increasing, and whole regions in our cities are unsafe after dark. Our courts' habit of freeing criminals has even brought the courts under attack. Reputable citizens are moving to the suburbs, and the regions they have left are labeled "ghettos." Police have to be stationed in a number of our racially integrated schools to keep the schools safe for both pupils and for teachers. Birth control, once a major eugenic remedy, is advocated by the United Nations and has deeply troubled even the Roman Catholic church. Recent discoveries of human chromosome irregularities have made it more difficult to ignore certain hereditary variables. The frequency of an extra Y-chromosome in certain types of male criminals has been found to be statistically significant. Perhaps the cause of the crimes they commit is not that they were "under-privileged" or that they were not loved enough when they were children.
>
> It may be that a more sophisticated and more modest eugenics will again become "In,"...

Zirkle's alarm at, at age 74, at the evolution of the modern world could not have been more apparent.

In any case, Zirkle loved Pickin's book. He also complemented the author for "bibliographical citations" that were "numerous" and "of great assistance to anyone who may wish to investigate the field." The index included references to Muller, Huxley, Haldane and, of course, Zirkle. The first was discussed in terms of his 1932

[66] For a notably different response to Picken's book see Mark H. Haller's review in *American Anthropologist* 71, no. 5 (1960): 993.

pro-eugenic anti-capitalist tirade which, the author dolefully concluded, did at least as much damage to Galtonian eugenics as the "fear" brought about by the Great Depression.[67]

As for Huxley, Pickens credited him with having pointed out that by "raising the average of desirable traits (greater physical vigor, resistance to specific diseases) a large burden of hereditary suffering would disappear from mankind's existence."[68] Haldane received more or less the same lukewarm treatment as Muller. He was counted among those (Ronald A. Fisher, Sewall Wright) whose work "displayed the mathematical consequences of diploid Mendelian heredity which eugenic reformers in the Galtonian tradition never imagined." That Haldane, Fisher and Wright were also jointly responsible for a far more monumental achievement—proving mathematically that Natural Selection could actually work in nature—seems not to have interested Pickens that much.

Zirkle was portrayed by Pickens in by far away the most positive light. Zirkle was quoted as though he had discovered that there may be more than one gene responsible for a hereditary character Zirkle. This was the really damning flaw in Picken's book, the weakness that showed the whole work for what it was. Why would you quote a botanist as though he were an expert geneticist?

Not everyone was as sanguine about Picken's book as Zirkle. Mark H. Haller, writing for the *American Anthropologist*, said that not only was the book poorly written, but the author was obviously unaware of the latest research published on the "race concept" in anthropology. Haller concluded:

> This is a book that, unfortunately, falls short of the promise inherent in the subject matter and the author's knowledge of published sources.[69]

With that in mind, one concluding remark. Though Dunn and Dobzhansky were not cited in Picken's book, Zirkle wrote about them in his. When discussing the value of eugenic selection for eliminating genetic "defects" such as albinism or blindness—an idea Dunn and Dobzhansky had described in *Heredity, Race and Society* as a waste of time—Zirkle asked whether it might not be worth pursuing afterall.

> What about the feeble-minded? Some 6 percent of our adult population has an I.Q. of 69 or less. What could eugenics selection do to reduce the number of dim wits? The answer depends, of course, on the genetic formulae which are responsible for human stupidity. But the matter at least deserves discussion in any popular treatment of eugenic selection.[70]

---

[67] Ibid, pp. 204–5.

[68] Ibid, p. 211.

[69] Mark H. Haller's review in *American Anthropologist* 71, no. 5 (1960): 993.

[70] Zirkle, *Evolution, Marxian Biology, and the Social Scene*, pp. 491–2.

## 6.6 Race and Eugenics

Clearly the reason that neither Dunn nor Dobzhansky were cited as important sources on eugenics in Pickens book is that their collective efforts, since the publication of *Heredity, Race and Society*, had been in the opposite direction. It should also come as no shock that they would have serious issues with Zirkle's *Evolution, Marxian Biology, and the Social Scene*. Dobzhansky reviewed it in *Science*, and Dunn in the *American Journal of Human Genetics*.[71] The latter sent Zirkle a copy of his review ahead of time. It is highly unlikely Zirkle was pleased by what he read.

Dunn essentially pointed out the obvious: The purpose of Zirkle's book was not to document the (*sic*) damage done to biology by Marxist philosophy, it was to imply that anyone who questioned eugenics or was a Lamarckist was also a communist. Dunn kindly avoided delving too deeply into the numerous scientific errors, misinterpretations and inaccuracies in Zirkle's book, but did wonder aloud why someone who—as evidenced by his published record as a historian of science—was so well-versed in the longer history behind ideas like Lamarckian heredity, would be so unable to view the situation in Soviet biology with a proper sense of perspective. Dunn's conclusion reflected an interpretation of the Lysenko affair that would soon become dominant.[72]

> In respect to biology in the Soviet Union, the chief effect of what Zirkle calls Marxian Biology—namely the suppression of genetics—was the result of a political decision of the Central Committee of the U.S.S.R. Communist Party and the part that Marxism played in that decision is nowhere made clear. This reviewer would hazard the guess, that the chief historical importance of that decision will eventually be found to lie in the assertion of dominance of political over scientific authority. That is an issue which every society of the future must face—and it merits more sober study than it has received in this book.[73]

In other words, Marxism was irrelevant.

Evidence that the disagreements between Lysenko's critics were becoming roughly comparable to the division that had once separated them all from Lysenko is even more apparent in Dobzhansky's review. He began,

> On 14 December 1958 *Pravda* denounced the author (Conway Zirkle) and the reviewer (Theodosius Dobzhansky) of this book as "reactionaries," because of their opposition to Lysenko's perversion of science. However, the two "fellow-reactionaries" find themselves at odds about several matters that are discussed in this book.[74]

---

[71] Theodosius Dobzhansky, "Evolution, Marxian Biology and the Social Scene," *Science* 129, no. 3361 (1959): 1479–80; L.C. Dunn, "Evolution, Marxian Biology, and the Social Scene," *American Journal of Human Genetics* 11, no. 4 (1959): 385–6.

[72] See Joravsky, *The Rise and Fall of T.D. Lysenko* and Graham, *Science and Philosophy in the Soviet Union.*

[73] Dunn, "Evolution, Marxian Biology, and the Social Scene," p. 386.

[74] Dobzhansky, "Evolution, Marxian Biology, and the Social Scene," p. 1479.

Clearly, the group of contributors to *Death of a Science in Russia* could never have agreed on an interpretation of the Lysenko controversy 10 years later. Dobzhansky, like Dunn, sniffed out Zirkle's deliberate ignorance in portraying Lamarckian heredity as part of some sort of "communist plot" in biology.

> Because he published in 1946 an excellent history of the idea of the inheritance of acquired characters, nobody is better qualified than Zirkle to know that, before the advent of modern genetics, inheritance of acquired characters was an admissible working hypothesis, and that at present the belief in such inheritance is an identification not so much of Marxism as of simple ignorance of elementary biology.[75]

Dobzhansky also recognized that Zirkle's goal in fomenting fears that Marxian biologists were hiding—as he put it—"under every laboratory bench," was to intimidate opponents of eugenics, as well as anyone who did not participate in condemning Lamarckism as an "enemy" doctrine in biology. Even more upsetting, by citing then disputing the passage from *Heredity, Race and Society*, Zirkle had basically lumped Dobzhansky in with other "Marxian biologists." Though Dobzhansky would later say that such ironies—accused by communists of being a racist and accused by racists of being a communist—were oddly reassuring, at the time they incensed him.

As mentioned, Dunn sent Zirkle a copy of his review in advance, and the latter's response was curious.

> Dear Professor Dunn:
>
> Thanks for the copy of your review. I would like to comment on only one of its statements, i.e., that I had the objective of persuasion. Now I had a number of objectives in writing the book and even included some experiments in it, but persuasion was not among them. I sought not to persuade but to annoy. This might seem to be pure frivolity on my part, but I assure you that it is not. I even took precautions not to convert anyone. You do not convert a man by attacking his religion head on, and in a manner that is mildly offensive.
>
> Most historians of science are convinced that the periodic changes that occur in the orientation of scientists are not due to any conversion of any scientist, but to the replacement of one generation of scientists by another. What has fascinated me in this is that often the changes occur in the complete absence of argument. One generation succeeds the other apparently without having made any intellectual contact. I cannot go into the whole problem here, but can only say that I hoped to stimulate some of my contemporaries into giving me data that, historically, is very scanty.
>
> I may have ultimately to ask forgiveness from a great many of my friends.
>
> Very truly yours,
> Conway Zirkle[76]

[75] Ibid.

[76] Correspondence, Conway Zirkle to L.C. Dunn, September 15, 1959. B: D917 Dunn. Zirkle, Conway.

Zirkle also corresponded with Dobzhansky on his motives for writing *Evolution, Marxian Biology, and the Social Scene*. Zirkle wrote that his goal was "to encourage Lysenko and co., at least temporarily. At the present state of the Cold War I would like to retard Soviet Agriculture and Medicine…" Dobzhansky wondered—in a letter to Isadore Michael Lerner, whether or not "such a Machiavellian behavior" was "compatible with ethics to which scientists supposedly adhere?"[77]

Whatever Zirkle might have been up to with his "experiments," it is apparent that by this time his behavior made him part of a group who had discovered the utility of claiming attitudes towards race and biology were indicators of support for communism. A helpful outline of this argument, along with recurrent use of the above-mentioned "equalitarianism", appears in a book Dobzhansky reviewed for the *Journal of Heredity* a few years after his review of Zirkle—*Race and Reason: A Yankee View* by *Carleton Putnam*.[78] Evidence that the ideas expressed therein were not aberrations, but part of the mainstream of American thought, appears in the form of an index card that came with every copy of the first edition. It read: "Because I believe this book to be of exceptional importance, I urge it upon your immediate attention." The card was signed by Senator Harry Byrd of Virginia.

The detail and complexity of Putnam's argument is both frightening and impressive. He inserted "equalitarianism" (i.e., belief in racial equality) as a step before "socialism" in Marx's classic formula of evolution to communism. He said the doctrine was launched by East European immigrants to the United States in the first decades of the twentieth century who, "smarting" from the fact that they had not been immediately accepted into American society, took over the universities where they could brainwash the current generation of college students into believing "negroes" were not inferior.

Putnam's descriptions of this plot carries overtones of a Cold War spy novel. In one passage he wrote,

> By mail, by telephone, and finally by personal visits, North and South, I found professional scientists aplenty who saw what I saw. And I discovered something else. One prize-winning Northern scientist whom I visited at his home in a Northern city asked me, after I had been seated a few minutes in his living room, whether I was sure I had not been followed. Another disclosed in the privacy of his study that he had evidence he was being checked by mulattoes at his lectures. All, when first approached, were hesitant, withdrawn and fearful, and the reason was not far to seek. Their employers on whom their livelihood depended—the universities, the museums, the foundations—were either controlled by equalitarians or were intimidated by the race taboo. The scientists whom these institutions employed, if they were ever to hint at the truth, must do so deviously, under wraps over wraps, half seeming to say the opposite.[79]

[77] Correspondence, Theodosius Dobzhansky to Isadore Michael Lerner, August 11, 1959. B: L563 Lerner. Dobzhansky, Theodosius #3, 1958–1959. The American Philosophical Society.

[78] Carleton Putnam, *Race and Reason: A Yankee View* (Washington, D.C.: Public Affairs Press, 1961).

[79] Ibid, p. 19.

According to Dobzhansky, *Race and Reason* was not only a best-seller ("or at least a best-giver") in the South, it was assigned reading at high schools and colleges in states like Louisiana, Mississippi, Alabama and South Carolina.[80] Perhaps predictably, Dobzhansky's efforts—as a scientist—on race were at times received in a manner similar to the response Muller got for his attacks on Lysenko in the *Saturday Review of Literature*.

For example, in a *New York Times* article covering the Supreme Court decision that a Florida law barring "Negro-White" cohabitation was un-Constitutional, the author declared that many anthropologists believed that "Negro-White fusion" had a "deteriorating effect on both races." Over the course of a series of letters to the editor that followed, arguing over whether or not this was, in fact, the dominant view among anthropologists, *Race and Reason* was cited as evidence that it was. Dobzhansky's negative review of the book was then trotted out to undermine Putnam's work, after which Dobzhansky was in turn derided as merely an "authority" on "fruit flies," not race.[81]

Speaking of Muller, though race was not an issue that aroused in him great concern, eugenics remained an obsession. In their later years he and Huxley attempted to put together a project of eutelegenesis, an effort only hobbled by the biological reality of old age.[82] Muller had obviously long given up on his view that socialism offered the best system of social and economic organization. In fact, the centerpiece of Muller and Huxley's project was, ironically, the primary institution of capitalism: a (sperm) bank. Also, a large portion of Muller's eugenic views was occupied by his concerns about the "genetic load" of negative mutations increasing in the human population, a problem exacerbated by atomic testing and the unchecked use of X-rays in medical practice.[83]

It is an incident that took place thanks to these fears which brings us full circle on Muller. On December 3, 1954, an article in the *New York Times* quoted Federal

[80] The Reminiscences of Theodosius Dobzhansky, pp. 451–2.

[81] See "Rejection of 'Color' as the Test of a Crime," *New York Times*, December 8, 1964; "Letters to the Times: Negro-White Fusion," *New York Times*, December 15, 1964; "Letters to the Times: Negro-White Fusion," *New York Times*, January 2, 1965; "Letters to the Times: Negro-White Fusion," *New York Times*, January 15, 1965; "Letters to the Times: Negro-White Fusion," *New York Times*, February 20, 1965.

[82] See for example, Correspondence, Julian Huxley to H.J. Muller, April 1, 1964. Series III, Box 34, folder 4. Julian Sorell Huxley – Papers, MS 50, Woodson Research Center, Fondren Library, Rice University; Correspondence, Eugene Kelly to H.J. Muller, May 20, 1963. Box 34, folder 5. Julian Sorell Huxley – Papers, MS 50, Woodson Research Center, Fondren Library, Rice University; Correspondence, H.J. Muller to Julian Huxley, May 23, 1963. Box 34, folder 5. Julian Sorell Huxley – Papers, MS 50, Woodson Research Center, Fondren Library, Rice University.

[83] Muller was far more concerned about the former than the latter—in large part because he believed it was necessary for the U.S. to test atomic weapons for use against the Soviet Union (Carlson, *Genes, Radiation and Society*, p. 354). Nevertheless, he was by no means, as indicated by the incident with the U.S. Atomic Energy Commission indicates, silent on this issue.

Civil Administrator, Val Peterson, advising anyone who lived within seven or eight miles of Washington, D.C. to build a bomb shelter in their backyard. Dr. Willard F. Libby of the Atomic Energy Commission (AEC) downplayed the dangers of fallout by saying the hazards "could be minimized by fairly simple methods," such as "staying indoors" or "taking cover behind a few feet of earth."[84] Reports like this convinced Muller that the U.S. government was deliberately misinforming the public on the hazards of radiation.[85]

In March, 1955, Muller received an invitation from the AEC to prepare a paper for the United Nations International Conference on the Peaceful Uses of Atomic Energy, scheduled for Geneva in August. Things proceeded smoothly until July when the AEC informed Muller that his paper had not been accepted by the UN selection committee for oral delivery. Muller was stunned. By this time he was already in Europe, having decided to use the trip as an opportunity to visit his wife's family in England and Germany.

Muller was next told he could do a five minute talk, but that offer too was withdrawn. Muller decided to attend the conference anyway as an observer. After several speakers favorably cited his research on radiation genetics, the vice-chair of the panel he was to have presented on suggested that the audience stand in tribute to Muller. They did, and then followed up with an extended ovation. Muller had become a martyr for his views on radiation. But why had the United Nations tried to silence him?

The answer is they had not. On September 17, *The Washington Post* published a story, "AEC Accused of Blocking A-Report," wherein the UN denied the AEC account.[86] As for the AEC's claim that the UN had been the ones to reject Muller's paper, the executive assistant to the Secretary General of the conference responded,

> The implication is totally false. Dr. George L. Weil (AEC technical director for the conference) wrote us June 30 that Dr. Muller would not be a member of the United States delegation and they did not want his paper ('How Radiation Changes the Genetic constitution') presented. As far as we're concerned, the paper was naturally of great interest. But if a country says the person who has written the paper should not be on the program, we have to do but agree.[87]

The AEC defended itself by claiming that Muller's intention to discuss the use of atomic weapons on Hiroshima and Nagasaki could lead to discussion of non-peaceful uses of atomic energy. This, according to the AEC, amounted to a violation of the "rules" since the conference was organized to discuss only peaceful uses of atomic energy.[88]

---

[84] "President Alerts Mayors on Attack," *New York Times*, December 3, 1954.

[85] Correspondence, H.J. Muller to M. Hersnat, December 11, 1954. Series: Correspondence 1910–1972, Box 1. Muller MSS, Lilly Library, Indiana University.

[86] "AEC Accused of Blocking A-Report," *The Washington Post and Times Herald*, September 17, 1955.

[87] Ibid.

[88] "AEC Explains Blocking of Muller's A-Report," *The Washington Post and Times Herald*, September 18, 1955.

In an editorial, "Muzzling Dissent," published two days later, the *Post* compared the AEC's actions to science policy in the Soviet Union:

> This double dealing involved much more than a discourtesy or injustice to a distinguished scientist. It involves a grievous blow to the prestige of the United States abroad, representing this country as one, like the Soviet Union, where scientific opinions are suppressed if they are at variance with official prejudices and policies. And, most serious of all, it involves the right of a self-governing society to learn what it needs to learn if it is to remain self-governing.[89]

Former AEC consultant Walter Lapp made the point more directly a few days later in an address urging President Eisenhower to be more candid about the size of the U.S. stockpile of atomic weapons. Lapp said the AEC's censorship of Muller in Geneva, "smacked of 'Lysenkoism—the weird Soviet policy of though control in genetics'."[90]

Eugene Rabinowitch, a former research associate on the Manhattan Project, defended Muller as well, and also found a way to invoke Lysenko. Rabinowitch explained that the AEC's defensiveness on the issue of radiation was due to the fact that "Communist and Communist-influenced opinion in Europe and Asia had seized upon" the warnings of biologists like Muller, even though "the official Soviet 'line'—as proclaimed by Lysenko but dismissed by all serious geneticists—denies the very existence of genes and maintains that hereditary properties can be changed by such means as diet."[91] In other words, now it was what Eisenhower had termed the "Military-Industrial Complex" that was persecuting Muller—not Stalin or Lysenko.

## 6.7 Old Age

In 1961 Dobzhansky visited J.B.S. Haldane in Calcutta. Like Haldane, Dobzhansky was interested in the caste system. During his visit Dobzhansky wrote most of a book he would call *Mankind Evolving*.[92] In a letter Dobzhansky sent to several friends describing the trip he wrote:

> I asked to be introduced to some students who were untouchables by origin, naturally without revealing to them the reason of my interest. They were like other students and seemingly were so treated by everybody. I have asked several professors about the academic records of the students of low-caste origins, and received a variety of answers; apparently nobody has made a real study of this matter, and it seems that public opinion would be

[89] "Muzzling Dissent," *The Washington Post and Times Herald*, September 20, 1955.

[90] "U.S. Bomb Pile Set at Several Tons of T.N.T. for Every Person on Earth," *The Washington Post and Times Herald*, September 21, 1955.

[91] "Banned Atomic Paper is Opposite of 'Alarmist'," *The Washington Post and Times Herald*, November 6, 1955.

[92] Dronamraju, "On Some Aspects of the Life and Work of John Burdon Sanderson Haldane, F.R.S., in India," p. 216.

> against undertaking such a study. Which is again a great pity, because the caste system in India is the grandest genetic experiment ever undertaken on human populations, and carried on for at least two millennia. Castes were Mendelian populations, isolated by social barriers, and also occupational groupings.[93]

Dobzhansky opened *Mankind Evolving* with a quote from a Dead Sea Scroll: "So walk I on uplands unbounded, and/know that there is hope … for that/which Thou didst mold out of dust/to have consort with things eternal."[94] He believed evolution was progressive—not only was it ongoing, but we are getting better. In this spirit Dobzhansky challenged not only the likes of Carleton Putnam, but also Nobel Prize-winning physicist William Shockley who maintained there was a "moral obligation" to "diagnose the origin of Negro I.Q. deficits," as well as scientists who argued that Cro-Magnon man was intellectually superior to modern humans.[95] Who we are is a product of our nature and our history, he believed, and Dobzhansky's world was thriving.

In 1958 Dobzhansky received the Kimber Genetics Award from the National Academy of Sciences for his work on the role of heredity and environment in evolution. The next year he was awarded a Guggenheim for his study of the fitness of races of the fly *Drosophila serrata* in different environments. In 1964 he received an honorary degree from Columbia, and a National Medal of Science. The next year he received a National Book Award for another work, *Heredity and the Nature of Man*. Dobzhansky appeared on television talk shows, broadcast at the same time as the Million Dollar Movie and Wide World of Sports. By the late 1960s his Russian accent and craggy features had become familiar to scholars all over the world. He won another National Book Award and more honorary degrees from Berkeley and the University of Padua. The Museum of Natural History in New York City awarded him, along with three astronauts from the Apollo 9 space mission, a gold medal. He was one of the world's foremost authorities on population genetics—linguist, horseman, explorer, world traveler, author, philosopher, and eloquent spokesman for science.[96]

Meanwhile for Dunn things had gone along differently. Dobzhansky would later say that old age brought Dunn disappointment from all sides.

---

[93] Correspondence, Th. Dobzhansky to I.M. Lerner, January 25, 1960. B: L 563 Lerner. Dobzhansky, Theodosius. Circular Letters—Travel #3. The American Philosophical Society.

[94] Theodosius Dobzhansky, *Mankind Evolving: The Evolution of the Human Species*, Mrs. Hepsa Ely Silliman Memorial Lectures (New Haven, CT and London: Yale University Press, 1962).

[95] William Shockley, "Models, Mathematics, and the Moral Obligation to Diagnose the Origin of Negro I.Q. Deficits," *Review of Educational Research* 41, no. 4 (1961): 369–77; Walter Sullivan, "Was the Caveman Smarter Than We Are?" *New York Times*, January 24, 1971.

[96] "Research Grants Made," *New York Times*, July 13, 1959, 50; "Academy Cites Seven for Science Work," *New York Times*, April 28, 1958; "Guggenheim Fund Grants $1,500,000," *New York Times*, April 20, 1959; "Television," *New York Times*, September 17, 1966; "Dobzhansky Gets Degree," *New York Times*, April 4, 1968; "Geneticist Reports Seeing Start of a New Species," *New York Times*, March 13, 1967; "Museum Will Close for Centennial Day," *New York Times*, April 6, 1969.

> In the early 40s, when you came to see Dunn, probably more than fifty percent of the time, there were one or more visitors sitting in his second little room, waiting for Dunn to see them, and another visitor was sitting in his office and talking. That took a lot of his time.[97]

Dobzhansky summed it up in his oral history by saying Dunn was a "political liberal" whose name appeared on countless letters, petitions, protests and so forth—most of which Dobzhansky himself would never have signed. It could be said Dunn was a bit "indiscriminate" in what he put his name to. He was also an active member of many organizations—friends of this and that: "As a matter of fact he spent—I think it can be freely said—too much of his time in those things." According to Dobzhansky, Dunn sacrificed his career for social causes: "I think Dunn would be in agreement if he heard me say that…"[98]

Dobzhansky added,

> He always felt, and I think quite justly, that he was on the side of the angels, and it probably was difficult for him to comprehend, or at least to be reconciled with, the fact that when he was denounced as, if not a criminal, at least some sort of an undesirable person…[99]

Dobzhansky also believed there was a "tragic situation" which could not be avoided when one spoke of Dunn, his oldest son Stephen, whom Dobzhansky described as: "A spastic, and a very bad spastic."[100] However Stephen had a genius I.Q. and received his PhD in anthropology from Columbia despite the fact that he was unable to write. According to Dobzhansky, Stephen's problems occupied Dunn's mind his entire life and depressed him continually. Dunn wondered what would happen once he was no longer there to help his son. Dobzhansky said the example of Stephen Dunn was "something that gives one pause": "As a matter of fact, this situation is of the kind which makes one immediately think of euthanasia."[101] Stephen Dunn, to Dobzhansky, was a "philosophical problem which is not easy to solve," someone who would be better off dead but had "adjusted to his misery."[102]

## 6.8 Dunn

How had Dunn adjusted? For one thing, he deeply regretted ever having participated in *Death of a Science in Russia*. Dunn's comments in his oral history make this clear, and also show how the timeline of events in ones' life may get muddled, or rearranged, as one grows older. He said,

[97] The Reminiscences of Theodosius Dobzhansky, p. 463.

[98] Ibid.

[99] Ibid, p. 467.

[100] Ibid, p. 464.

[101] Ibid, p. 466.

[102] Ibid.

> I think the way to judge the positions of Americans—such as myself, Muller, Sax—is to take a book in which many of these opinions and positions were gathered together, published by Zirkle in'49, called Death of a Science in Russia. Most people who had published something on this acceded to Zikle's request to reprint, because he reprinted some original documents in his book. But his attitude was so clear, as expressed in the title, that I for one hesitated before I agreed to appear within his covers. And when I reviewed the book, I more or less said as much.[103]

In fact, Dunn never reviewed *Death of a Science in Russia*. He was conflating it with his review of *Evolution, Marxian Biology, and the Social Scene*. Moreover, if Dunn had written a review of the former publication saying he regretted his participation, he would have been excoriated. Reviews of *Death of a Science in Russia* when it appeared betray zombie-like affirmations of loyalty comparable to the reactions of geneticists at the VASKhNIL session.[104] Granted, the shock of the attack on genetics was still fresh when Zirkle published. But at least geneticists in the USSR, Poland, Hungary, Czechoslovakia etc. had the excuse that dissenting from the "party line" could mean imprisonment, ostracism or, at best, damage to their careers. U.S. and British biologists—in theory—were free from such concerns.

In any case, Dunn's thoughts on the book, 12 years later, are of interest. He said,

> It stirred up tremendous emotional responses. I think Muller corrected his position later, but at that time he was extremely emotional about the whole matter, and I thought missed some scientific judgments involved. Zirkle perhaps wasn't entitled to a strong scientific position, since he hadn't worked very actively in genetics. He was very well qualified to judge the state of development of controversy about the inheritance of acquired characters, because he'd done a good historical job on that, but his political position was tainted by what we would call reaction—I'm sure that's his normal position.
>
> ... Now, the emotions that were kicked up were partly indignation, but partly it was because these same people saw Reds under other beds besides those inhabited by geneticists. Many of the criticisms in fact had very little to do with the scientific merits of either side of the controversy. They were based on emotion and defense of political positions, which was the very thing they were criticizing in the people taking part in the debates.[105]

"Reds under other beds," or under "every laboratory bench" as Dobzhansky put it. That was where people like Zirkle placed people like Dunn. For this reason it is worth reviewing Dunn's assessment of this whole (Lysenko) affair. Eugenics: The problem was exaggeration.

> Davenport understood Mendelism perfectly well, but he gave it an enormously wide extension. He even wrote a paper on the Mendelian inheritance of the sea-roving instinct! I think Thalassophilia, or some nice Greek name like that, was what he called it. This was utter nonsense. It simply traced the descent of a number of sea captain families from New England and the East coast, and showed that a succession of descendants in that family had gone to sea. That's all it showed.[106]

---

[103] The Reminiscences of L.C. Dunn, p. 778.

[104] See, for example, Bentley Glass, *The Quarterly Review of Biology* 27, no. 1 (1952): 60–1; Karl Sax, *Isis* 41, no. 2 (1950): 238–9.

[105] The Reminiscences of L.C. Dunn, p. 779.

[106] The Reminiscences of L.C. Dunn, pp. 163–4.

Michurin: Once again, exaggeration.

> The Soviet geneticists had successes and they had failures. One of those successes was precisely Michurin. … he did some very good things, especially with fruit crops. There's no belittling the work that Michurin did—then it got overblown on the scientific side—but on the side of practical agriculture, it didn't.[107]

Lysenko: Dunn was too open-minded.

> It happens that some part of my attitude—namely, being patient and ready and willing not to only read it but to get some of it published—was due to the fact that one of the neglected areas on genetics was precisely in the way in which genes influence development. And this was one of the main arguments of Lysenko, and one of his accusations against his orthodox genetical colleagues was that they too had neglected the study of development, and he was undertaking this study, which was true.
>
> Now, he gave no new insights, that's true. But he did point out weaknesses in the development of the field until that time. And I think this attitude made me receptive because my interests were then mostly in development. Now, other people—well, take Muller. …

And what about Muller?

> Muller couldn't see that I was in no way taking Lysenko's part by taking my own position, because this was whole hog or none, as far as he was concerned. As I say, I think he's amended that position since. In fact, there are ways of being a little open-minded, and I think it is that that I was accused of. My mind, according to some of my correspondents, should have been made up a long time before to complete rejection and denunciation of Lysenko and all his works and everybody that stood with him.[108]

It was this sense of being misunderstood, by Muller, the State Department, his colleagues at Columbia, that made Dunn empathize with his Soviet counterparts. As Dunn noted, when he was being investigated by the State Department, the implication was that a scientist's place was in the lab, not "messing around" with foreigners. He was struck by the irony that this was exactly the Soviet position: Scientists should stay out of politics.[109]

Dunn had still been a student at the time of the Bolshevik Revolution and become fascinated because, as he put it, "political developments eventually spread round the world in one form or another, and what happens in one place at one time may stand a good chance of happening in another place at another time."[110] Dunn perhaps best expressed his enthusiasm when he wrote, in the essay ultimately published in *Death of a Science in Russia*, that, "Whereas Westerners were inclined to go in through the traditional front door, our Soviet colleagues seemed at times to break in through the back door or even to come up through the floor." For Dunn the USSR provided a natural experiment for observing how modernity might otherwise be. The fact that he did not view the United States and the Soviet Union as opposites, but rather as

---

107 Ibid, p. 690.

108 Ibid, pp. 780–2.

109 The Reminiscences of L.C. Dunn, pp. 797–8.

110 Ibid, p. 797.

societies using alternate methods to achieve parallel goals, with much to learn from one another, must have made the persecution he experienced for his Russian sympathies all the more baffling.

When Dunn retired there was no kind of dinner, no acknowledgement from his colleagues. This, according to Dobzhansky, was very unusual. The year before Dobzhansky had written to Dunn that he was very concerned.

> It is quite clear to me that you can't continue in your present state of mind. It will destroy all the joy of work, the very reason of work and of life and will end by affecting not only your pleasure in it but the work itself. And this must not happen now, at the height of your powers. I can not sit by and see all this creative energy frustrated and I shall do anything in my power to prevent it. ... But please don't sit and torment yourself in this depression.[111]

Dunn was not only despondent over having suffered for his political activism, but also because he came to feel alienated from the next generation of politically active scientists. Before the Second World War Dunn always seemed to be on his way to and from meetings so that his time in the laboratory and with students—aside from lectures—became constricted to early mornings, late nights and weekends. "I was extremely conscientious about this," he said, "always feeling that if I were slack, the wheels of progress would slow down all along the line and many people would be affected."[112]

This accentuated a sense of responsibility based in the fact that he had always felt like the member of a "far-left minority." There were so few of them it seemed he must work that much harder to make up for it. As a result he belonged to many quasi-political causes and rarely refused appeals. "This made for a very busy, very active life," he said, "but since my scientific work was making some progress, I had no regrets."[113]

After the war, however, Dunn began to feel a certain sense of futility, and an inability to participate in the rapid give and take of debate in science and politics. To him the physicists, chemists and engineers who had played a vital role in the atomic age seemed like, "a new kind of person, scientists who had become acutely conscious of the social and political forces by which the world was being changed so rapidly, yet recognizing that the fundamental changes had been set in motion by science," for which they were responsible. This is how Dunn felt too, yet he found himself unable to act as they did. Maybe it was the McCarthy-era, the loss of his passport—but how we behave, he said, "is so much influenced by personal and private changes which are often not identified that causal analysis is never exact or complete."[114]

Dunn said scientific ideas are judged by the future, never by the past. What an idea becomes means more than how it got there. The best ideas are those that are not

---

[111] Correspondence, Th. Dobzhansky to L.C. Dunn, July 13, 1961. B: D65 Dobzhansky Papers. Dunn, Leslie Clarence #2. The American Philosophical Society.

[112] The Reminiscences of L.C. Dunn, p. 1026.

[113] Ibid.

[114] The Reminiscences of L.C. Dunn, pp. 1024, 1027.

idle, when you can say, "that's a good idea because I know what to do about it."[115] For Dunn Lysenko was appealing because he opposed views Dunn found repugnant. Regardless of the fact that cerebral palsy is not a genetic condition, Dunn must have known what would have happened to his son under even the mildest enforcements of eugenics. The belief in the value of his own child's life was an important idea—no matter "how it got there."

***

This chapter begins with an anecdote about the application of Lysenko's theories to animal breeding. Lysenko studiously avoided applying his ideas to humans, a caution Lepeshinskaya ignored. But was this notion not implied? If so, what were the implications?

How would you compare Dunn's problems with the State Department to Muller's interview with HUAC? Do you think Muller would have said Dunn was an "egghead?" To what extent was Muller talking about himself?

What about Zirkle's concerns about efforts to rehabilitate Lamarckism. Were Lysenko's theories really "Lamarckian?" What are the similarities and differences? Also, how does this fit in with the question, raised at the end of Chap. 4, of whether the Lysenko controversy caused prejudice towards theories of heredity which emphasized environmental factors in evolution during the second half of the twentieth century?

As for Khrushchev's interest in Roswell Garst—how was this ironic? How was it similar to Lenin's interest in Burbank?

What about Huxley's debate in Karachi, and his over-simplified version of what took place? Considering that this debate took place one year before the 1955 Bandung Conference in Indonesia, how do you think events like this might have influenced the formation of the non-aligned movement?

Related to this was Haldane's self-imposed exile in India. Leaving aside his primary and secondary motivations for doing so, what do you make of the anecdote that appeared in Reuters the same day, "Haldane His Own Rabbit?" Why was the press so interested in the fact that Haldane experimented on himself?

What about the claim made after the launch of Sputnik that Lysenko had lulled the West into thinking Soviet science and technology was less advanced. While it would be ironic if this could indeed be counted among deceptions linked to Lysenko, do you think it was true?

Re-read footnote 53. How is this anecdote ironic? Does it surprise you?

To what extent did the discovery of the chemical structure of DNA play a role in Lysenko's downfall? On the one hand, the timeline of when Lysenko suffers his first challenge and James Watson and Francis Crick publish their work seem to indicate this. But is it that simple?

[115] The Reminiscences of L.C. Dunn, pp. 1045–8.

Compare Zirkle's "Diagram of Interaction of Cultural and Biological Evolution" on p. 154, with this diagram from Lenin's *Materialism and Empiriocriticism*:[116]

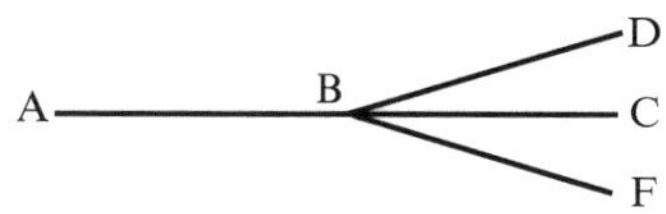


This diagram is intended to mock empiricist philosophy.[117] How is Zirkle and Lenin's use of diagrams similar or different? How would the enlistment of something that looks "mathematical" (in Zirkle's case) or the debunking of something that looks "mathematical" (in Lenin's case) serve analogous purposes?

Referring back to Zirkle's neologism, "Lysenkoids," what word does it remind you of? What does it imply technologically? Is there anything significant about the evolution of the term "Lysenkoist" to "Lysenkoid?"

As for Zirkle's attitude towards Lamarck, how might the concept of "whiggish" behavior be considered biologically? Do we prefer to feel superior to our ancestors? Do we hope that human evolution will be progressive, i.e., that our successors will have reason to consider us "less advanced?"

What about Zirkle's review of *Eugenics and the Progressives*. What elements of Zirkle's attitude towards the younger generation make sense to you? How does this square with his comment to Dunn that, "Most historians of science are convinced that the periodic changes that occur in the orientation of scientists are not due to any conversion of any scientist, but to the replacement of one generation of scientists by another?" What other philosophies of scientific evolution does this remind you of?

Also, how is Zirkle's view, ironically, reminiscent of what Safanov wrote, quoted in the discussion questions section at the end of Chap. 3 (the "classics of science" have always been written in "rebellion against what in their time was regarded as incontrovertible knowledge")? How can it be related to the episode recounted at the beginning of Chap. 4 where, as Lysenko walked through a crowd on his way to give a speech he shouted—"Aha! You came to relearn?"—to older scientists waiting outside? How does it reflect the notion of evolution?

Again, with regard to Zirkle's attitude towards 60s youth, what does he mean when he says: "Birth control, once a major eugenic remedy, is advocated by the United Nations and has deeply troubled even the Roman Catholic church?" What do

---

[116] V.I. Lenin, *Materialism and Empirio-Criticism* (Moscow, Russia: Foreign Languages Publishing House), p. 163.

[117] The original version appears in Joseph Petzoldt, *Einfurhrung in die Philosophie der reinen Erfahrung* (Leipzig, B.G. Teubner, 1900–1904), p. 37. Lenin's is a simplified version of Petzoldt's diagram. It may be worth having a look at the original to think more deeply about what Lenin was doing with his diagram. See also Hudson and Richens reference to the book in *The New Genetics*, p. 23.

you make of this—coming from Zirkle—as an argument that eugenic ideas are once again "in?" What do you think he thought of the sexual revolution?

What about Pickens quoting Zirkle as an authority on genetics?

As for Haller's review of Pickens and his focus on sources, go through an article or book—historical, scientific, or both—that you enjoy. Is there anything apparent in the author's citation of sources?

Do you think the comparison between Lysenkoism in the USSR and Muller's treatment by the AEC are comparable? What are the virtues of this comparison? What are the flaws?

Going back to the uniformly positive reviews of *Death of a Science in Russia* when it appeared, were U.S. and British biologists free from concerns about following a "party line?" What evidence from the events described in this book would you use as evidence for either side of the argument?

Do you agree with Dobzhansky's opinion that Zirkle's goal to deliberately undermine Soviet science was "Machiavellian?" If so, then under what circumstances would it not be?

Also, it is ironic that Muller had gone from thinking the Soviet Union was the only place a eugenics policy would work to attempting to found a sperm bank. But could it be said that Muller's scientific views ultimately transcended politics? What was the relationship?

Dobzhansky believed that the "caste system in India is the grandest genetic experiment ever?" Why do you think he was baffled by the fact that no one had "made a real study of this matter?" Do you find that surprising?

What about Dunn's self-assessment that he had been "too open-minded?" Do you think his analogies about U.S. and Soviet scientists' relationships with their respective governments were correct? If so, in what ways?

And then there is Dobzhansky's assessment of Dunn's career. Do you agree or disagree? Do his remarks about Stephen Dunn surprise you?

Look at the "Who's Who? What's What?" questions from the *New York Times* I list at the start of the Epilogue below. Why—aside from the one on Lysenko—do you think I included them? How are the succession of Democratic presidents in the White House, and the governments of South Vietnam and Rhodesia, related to the Lysenko controversy? Can they be?

In the Epilogue section you will also read a comparison between Rasputin and Lysenko. Do you think there is anything to it? Were they alike? How?

Pay close attention to Loren Graham's account of his meeting with Lysenko. What do you think of Lysenko's comparison of himself to another persecuted minority? Explain.

Finally, could the entire Cold War be explained in terms of biology?

# Epilogue: "Nikita and I have our difficulties..."

## Who's Who? NEWS QUESTIONS What's What?

1. Lyndon B. Johnson last week became the sixth Democrat elected President since the Civil War. Name the other five.

...

8. With reference to the new government formed in South Vietnam, pair the following—Nguyen Khanh, Tran Van Huong, Phan Khac Suu—with their posts: Chief of State, Premier, head of the armed forces.

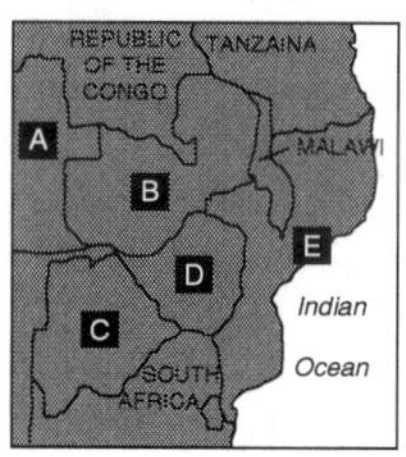


...

9. In a referendum last week the voters of Rhodesia—mostly whites—voted in favor of independence for their country from Britain. Identify Rhodesia on the map.

...

10. There were indications last week that among those who may go into eclipse in the Soviet Union as a result of the downfall of Nikita Khrushchev is Trofim D. Lysenko. Is he known for his work in disarmament, genetics, or Marxist theory?

—*New York Times*, November 8, 1964[1]

[1] "Fifteen News Questions," *New York Times*, November 8, 1964.

W. deJong-Lambert, *The Cold War Politics of Genetic Research: An Introduction to the Lysenko Affair*, Archimedes 24, DOI 10.1007/978-94-007-2840-0,

## *Rasputin*

Lysenko was born two years before the start of the twentieth century, during the reign of the last Romanov, Tsar Nicholas II. Few people in Lysenko's village probably remembered serfdom, though it had only recently been abolished. The life of the average Russian lasted less than 40 years. Time meant nothing. The workday could last from sunrise to sunset, or only a few hours, depending upon what needed to be done. In the winter when the ground was hard and the weather was cruel days might be spent asleep on the stove. Few people could read, or needed to. Life was unpredictable and full of magic. Fairies, water nymphs, demons and devils filled the forests, and everyone feared the evil eye. The season to reap or sow was determined by signs like, "when the tree gets dressed," or a certain bird arriving in the village. Tools and techniques—plows or weeding—brought from the city, caused suspicion or ridicule; worse than evil was being laughed at.[2]

The Tsar ruled the world along with God, and would make their lives better if only jealous nobles did not keep the peasant's misery from his ears. The Royal Family's greatest weakness was the son Alexis. Alexis was afflicted with hemophilia, a hereditary blood disorder known as "the royal disease." The blood of hemophiliacs has problems clotting: A minor cut could mean bleeding to death, a bump or fall could cause permanent crippling. The disease spread among the monarchies of Europe as a result of intermarriage, a recessive trait passed along by the genes in Mendelian ratios. Doctors were helpless, and the Empress Alexandra spent hours a day praying. One day God brought her Rasputin.

Grigori Rasputin was a holy man from Siberia rumored to have miraculous powers of healing. He had eyes like a lynx which easily hypnotized. His hair was tangled, his beard was matted, and he ate soup with his hands. He was surrounded by rumors of nocturnal orgies in the forest and pagan ways of getting closer to God. Rasputin was an object of fascination in the dying days of the Empire because of his relationship with Alexandra. He was able to stop Alexis's bleeding.

As the Russian army fell apart during the First World War, Nicholas left for the front to take personal command of the troops. He left the Empress and Rasputin in charge. Rasputin's assassination by jealous ministers was a prolonged process consisting of poison wine, poison cakes, bullets to the back and head, and beating with a rubber club in the frozen snow. His unconscious body was then rolled up in a curtain, tied with rope and tossed into the hole of an ice-covered river, where he finally drowned.[3] Lysenko was often compared to Rasputin.[4]

The answer to number ten on the list of news questions in *The New York Times* appeared on page seven: "Genetics. The Lysenko school, which was made dogma

---

[2] Maurice Hindus, quoted in Ronald Grigor Suny, *The Soviet Experiment: Russia, the USSR and the Successor States* (New York and Oxford: Oxford University Press, 1998), p. 11.

[3] Description of Rasputin's death comes from Robert K. Massie, *Nicholas and Alexandra* (New York: Atheneum, 1967), pp. 353–63.

[4] See, among numerous other examples, Huxley, *Heredity East and West*, p. 101, cited above.

under Stalin after 1948, developed new theories of plant-breeding in conflict with classical genetic theory. Premier Khrushchev supported the Lysenko theories, which now appear to be out of official favor."[5] At this point it could be said that Lysenko had been tossed through the ice.

After Sputnik there was Gagarin, the first human in space. Khrushchev's boast: "We will bury you," seemed possible, even if it meant building a wall in Berlin to keep people from fleeing to the West. Stalin's body was removed from Lenin's mausoleum even as the cult of Khrushchev emerged with the requisite props of genius—published works, portraits, praise. But his success or failure was tied to agriculture, and Khrushchev had bound himself to Lysenko.

In January of 1961 Lysenko complained to Khrushchev that it had taken the Ministry of Agriculture six months to reply to a 20-page plan he had sent them on how to increase the butterfat content of Russian milk from 3.5% to 4.5%. Lysenko proposed crossing Jersey cattle with local breeds, but insisted that the males must be smaller than the females. If a large bull with a high butterfat genotype bred with a small cow the offspring would be small because the zygote would sense that a large calf would have difficulty emerging from a small cow. Thus, in order for the high butterfat trait to be transmitted the cow must be larger than the bull. "The zygote is no fool," Lysenko stated, and asserted following his program would mean an additional 357 ½ gallons of milk per cow each year in the Soviet Union. Khrushchev concurred by scolding "slow-reading bureaucrats."[6]

Lysenko's dizzying comeback was chronicled in the West. One of the most sinister influences of the Stalin-era—a man with burning, deep-set eyes who relentlessly indicted his opponents; only now the hair behind his distinguishing forelock was a little grayer. The story went that Lysenko had fallen into disfavor by opposing Khrushchev's corn program. After that he had rolled up his shirt sleeves and stumped the country, preaching the power of pragmatic farm practices. By trading in a genetic theory that fit Communist party doctrine for a focus on increasing agricultural production, he had re-emerged as the voice for those who preferred the muck of the barnyard and dirt of the fields to the sterile laboratory.[7]

But it did not last long. In the Ukrainian city of Ostrog a bronze sculpture of Lysenko and Stalin seated, facing one another in earnest consultation, was removed.[8] Not long thereafter Lysenko was forced again to resign as president of the

---

[5] "Answers to Questions on Page 7," *New York Times*, November 8, 1964.

[6] Soyfer, *Lysenko and the Tragedy of Soviet Science*, pp. 219–20; Medvedev, *The Rise and Fall of T.D. Lysenko*, pp. 187–90; "Khrushchev Scolds Farm Bureaucrats," *New York Times*, January 16, 1961.

[7] ; "Lysenko is Seen Regaining Power," *New York Times*, January 28, 1961; "On the Rise Again," *New York Times*, January 28, 1961; "Lysenko Stages Soviet Comeback," *New York Times*, August 13, 1961; "News Notes," *Science* 133, no. 3452 (1961): 568.

[8] Theodosius Dobzhansky, "Soviet Biology and the Powers that Were," *Science* 164, no. 3887 (1969): 1507; Barry Mendel Cohen, "The Rise and Fall of T.D. Lysenko: An Essay Review." B: L563 Lerner. Medvedev, Reviews and Miscellaneous Clippings #3. The American Philosophical Society.

All-Union Academy of Agricultural Sciences. The manuscript of a book which would later be published in the West as *The Rise and Fall of T.D. Lysenko* began circulating in the underground, and its author, Zhores Medvedev, co-authored an article in a Soviet literary journal entitled "Prospects of Soviet Genetics." These events were watched closely in the United States, and interpreted as a "renewed attack" on Lysenko.[9]

## *Some Endings*

On October 12, 1964 the Soviets launched Sunrise, with three Cosmonauts on board. The purpose of the mission was to conduct biological tests to determine the influence of space conditions on the human organism. Instead of spacesuits the Cosmonauts wore a light-weight costume made of steel-colored wool with white helmets. Sunrise returned to earth 24 hours later, passing across the Aral Sea just before it landed.[10] Though no one knew it yet, the Aral Sea had begun shrinking the previous year as water from the two rivers which fed it was continually siphoned off to grow cotton on the arid landscape of Central Asia. The port towns would become stranded as the shore receded, leaving behind rusting ships in a salty desert as mute testimony to the transformation of nature.

Khrushchev had been vacationing in the Crimea, where he took the opportunity to inspect the chickens at nearby farms. The month before he had announced a plan to increase Soviet egg production from less than 9 billion to 30 billion a year by 1970.[11] He was called back to Moscow, brought to the Kremlin, and forced to resign.

Reaction was relatively mild. A story in the *New York Times* described how on Wall Street a broker skipped lunch and remained at his desk smoking 14 filtered cigarettes, expecting a rush of calls from panicked customers—but almost none came.[12] A weight-lifter on the Soviet Olympic team in Tokyo expressed indifference, and the motion for a mistrial in a Russian espionage case in New Jersey was denied: "What would you have the court do," the judge asked, "postpone the trial until we have an era of stability and guaranteed peace?"[13] On the campaign trail

---

[9] Robert R. Herr, "Renewed Attack on Lysenko's Genetic Theories in the Soviet Union," *AIBS Bulletin* 13, no. 4 (1963): 50–1; Soyfer, *Lysenko and the Tragedy of Soviet Science*, pp. 272–7; "Lysenko Resigns Soviet Farm Post," *New York Times*, April 6, 1962.

[10] "Astronauts on T.V.," *New York Times*, October 13, 1961; "Soviet Spaceship is Landed Safely After Sixteen Circuits," *New York Times*, October 14, 1964.

[11] "Khrushchev Views Chickens," *New York Times*, October 5, 1964.

[12] "Wall Street is Wary in Face of News," *New York Times*, October 16, 1964.

[13] "Soviet Stars Unfazed by Khrushchev Ouster," *New York Times*, October 17, 1964; "Events in Moscow Echo at Spy Trial," *New York Times*, October 17, 1964.

Senator Barry Goldwater deleted several references to "good old Nikita" from a speech text.[14]

In Iowa, Roswell Garst told reporters he was "'mystified, confounded and confused' by Mr. Khrushchev's downfall." Regardless of changes in the Kremlin, he said: "The Russian people are damned anxious to eat better than they have. The new leaders will be under pressure to improve the Russian diet."[15]

In the Soviet Union the capital was reported to be quiet and deserted. A lone pedestrian strolling the streets was quoted: "How unexpected this is. And I thought he was going to meet the cosmonauts in Red Square when they came to Moscow."[16]

Lysenko was referred to as the most prominent casualty of the changes in the Soviet Union, next to Khrushchev's son-in-law. An irreversible campaign was now launched to break his iron grip on biology, as he was singled out for criticism in newspapers and on the radio. *Pravda* announced textbooks and curricula would be rewritten, a staff member at his own institute wrote in a newspaper that Lysenko's bulls were not as good as he had proclaimed, and the campaign was widened to remove the members of his personality cult at ministries and institutes. A few months later, at age 66, Lysenko was forced to resign his post as Director of the Institute of Genetics at the Academy of Sciences. A report was made public within the year describing his butterfat work as a "lie" and organo-mineral fertilizer claims as "misrepresented."[17]

Just before the fall of Khrushchev, Raissa Berg, who had accompanied Vavilov and Muller on their last night in Leningrad, gave a lecture at a high school in Novosibirsk. In her autobiography she later recounted how the older generation of biologists she spoke to had been forced to believe in Michurinism while the younger had never known anything different. The collective farms did not provide them with food and there was nothing to buy in the shops. They were allowed to keep one cow and received free electricity for one light bulb. A second was forbidden.

Her lecture was anti-Lysenko, but when Berg was done speaking an audience member asked what was the position of Michurinist biology on the questions she

[14] "Goldwater Changes Text," *New York Times*, October 16, 1964.

[15] "Iowan 'mystified' by Khrushchev's Fall," *New York Times*, October 17, 1964.

[16] "Moscow's Streets Are Virtually Deserted as 11-Year Khrushchev Era Ends," *New York Times*, October 16, 1964.

[17] E.W. Caspari and R.E. Marshak, "The Rise and Fall of Lysenko," *Science* 149, no. 3681 (1965): 275–8; Soyfer, *Lysenko and the Tragedy of Soviet Science*, pp. 280–8; Medvedev, *The Rise and Fall of T.D. Lysenko*, pp. 225–38; "Pravda Appears to Condemn Lysenko Genetics," *New York Times*, November 3, 1964; "Biology Teaching Decried in Soviet," *New York Times*, November 12, 1964; "Attack on Lysenko," *New York Times*, November 12, 1964; "Soviet Condemns Lysenko by Name," *New York Times*, November 11, 1964; "Attack on Lysenko Widened in Soviet," *New York Times*, November 20, 1964; "Russia: The Party Line," *New York Times*, November 22, 1964; "Colossal Harm is Laid to Lysenko," *New York Times*, November 25, 1964; "Russians Challenge Khrushchev's Ideas on Broad Front," *New York Times*, January 13, 1965; "Lysenko Dropped as Genetics Chief," *New York Times*, February 5, 1965; "Fraud Charge Against Lysenko is Publicized Widely in Soviet," *New York Times*, January 7, 1966.

had illuminated. There is no Michurinist biology she replied. They protested that 300 different kinds of fruit and berries had been derived by Michurin. Even though they might never have eaten an apple or be able to tell a peach from an apricot, they believed it. She wanted to tell them how apple vendors in the Leningrad markets would advertise—"These are real apples, not any of Michurin's graftings"—but then she realized she would be taking away their faith. The audience believed that even if they would never eat 300 sorts of exciting plants their children might, or maybe their grandchildren anyway.[18]

Over a year later it came out that 15 miles south, in an experimental school administered by the University, students had been learning the latest findings in modern genetics for quite a while. The school had been kept secret, since it was not under the jurisdiction of the Ministry of Public Education in Moscow. The parents of most of the students were teachers, engineers, physicians and other types of skilled workers. Only a few came from families of local government or Communist Party officials.[19] Though Lysenko had set out to remove "bourgeois elitism" from scientific practice he had in certain ways reinforced it. As Berg realized, Michurinism had not modernized the peasant, but become another sort of superstition—albeit a scientific one. The intelligentsia had not been persuaded, however, and quietly found a way to circumvent it.

In the fall of 1963 J.B.S. Haldane was diagnosed with cancer. In the hospital hooked up to tubes he wrote a letter to science fiction writer Arthur C. Clarke saying it was a "foretaste of the future," when science will sustain us with a variety of gadgets long after our natural body has wasted away. The BBC asked Haldane if he would be willing to tape a self-obituary from his hospital bed. He agreed but recovered well enough to do it in his old room at University College, London. Haldane appeared in flowing Indian robes, puffing away on a pipe.

"I am going to begin with a boast," he said, "I believe that I am one of the most influential people living today, though I haven't got a scrap of power. Let me explain. In 1932 I was the first person to estimate the rate of mutation of a human gene—and my estimate was not far out."

Going on to describe his other achievements, Haldane also took the time to praise Lysenko: "In my opinion Lysenko is a very fine biologist and some of his ideas are right."[20]

Less than a year later Haldane was dead. "The genetic heaven must be a place in which there is room for all sorts of people," he had once said, "each best at something or other."[21]

---

[18] Berg, *Acquired Traits*, pp. 340–1.

[19] "Siberian Experimental School Stresses Genetics," *New York Times*, November 27, 1964.

[20] "J.B.S. Haldane's self-obituary," *Listener*, December 10, 1964, pp. 934–5.

[21] "Prof. J.B.S. Haldane, 72, Dies; British Geneticist and Writer," *New York Times*, December 2, 1964.

## *The Rest*

In May, 1966 Conway Zirkle celebrated his 71st birthday and retired. He received numerous letters and telegrams in congratulations, including a handwritten missive from Lysenko:

> Moscow, 1.V.1966
>
> Dear Conway!
>
> You cannot imagine how much it burdens me to be unable to participate in the gala celebration honoring your long academic career. The whole world knows that it is primarily through your efforts and publications that I have gained an international reputation in science. In fact I have reached the stature and even the material mode of existence of a premier of this vast and rich country. While Nikita and I have our difficulties in sharing this one-room apartment, we none the less join in sending our heartiest felicitations.
>
> Your loyal comrade,
> Trofim Denisovich Lysenko[22]

In fact, it was a joke, written by a friend. Another friend wrote him a message from "Gregor Mendel" which read: "Congratulations Professor. I've always been proud you stuck to our side."[23]

In 1967 at age 76 H.J. Muller died. He had suffered a heart attack a few years before and was taken to the hospital on his birthday, suffering from kidney failure. After a few months Muller realized he was not going to recover, and decided to stop eating.[24]

That summer there was a meeting of the International Astronomical Union in Prague. The members stood before a gigantic map of the far side of the moon, debating its accuracy. The United States cited errors in previous Soviet maps, and suggested waiting until more data was available from the American Orbiter 5 space craft before assigning names to geographic features. To everyone's surprise the Soviets agreed, and also pointed out that they had chosen to name a part of the moon "Mendel." This was taken as a friendly signal in the West, because the name was not Russian. A year and a half later a 70-mile wide crater was named "Vavilov."[25]

Lysenko was banished to his model farm in the Lenin Hills where he issued periodic reports which were ignored. He remained elusive to his critics in the West—his status and health a continual mystery. In 1969 professor Martin McCauley from the School of Slavonic Studies at the University of London asked for official permission to see Lysenko. The request was turned down because supposedly he was dying of

---

[22] B: Z67 Conway Zirkle Papers. Lysenko, Trofim Denisovich. The American Philosophical Society.

[23] B: Z67 Conway Zirkle Papers. Birthday Greetings 1966. The American Philosophical Society.

[24] Carlson, *Genes, Radiation, and Society*, p. 421.

[25] "Quarrel Averted on Lunar Names," *New York Times*, August 25, 1967; "Moon Crater Named for Once-Disgraced Soviet Geneticist and Brother," *New York Times*, December 3, 1968.

diabetes. However, according to McCauley, it was also rumored among students that, "Lysenko is on the up once again."[26]

Two years later Loren Graham, whose previous attempts to meet with Lysenko were unsuccessful, finally ran into him in the lunchroom of the House of Scientists in Moscow. Lysenko, though still a full member of the Academy of Sciences, was ostracized from the scientific community, and eating lunch alone. Graham sat down next to Lysenko, introduced himself, and began discussing Lysenko's role in the death and imprisonment of Soviet geneticists like Vavilov. According to Graham, "Lysenko abruptly stood up and left the table," but then 10 min later—much to Graham's astonishment—returned and sat down next to him. He said,

> You think I am a part of the Soviet oppressive system. But I have always been an outsider. I came from a simple peasant family, and in my professional development I soon encountered the prejudices of the upper classes. Vavilov came from a wealthy family and knew many foreign languages. When I was a boy I worked barefoot in the fields and I never had the advantage of a proper education. Most of the prominent geneticists of the 1920s were like Vavilov. They did not want to make room for a simple peasant like me. I had to fight to be recognized. My knowledge came from working in the fields. Their knowledge came from books, and was often mistaken.
>
> And, once again, I am now an outsider. Why do you think I was sitting alone here at this table when you came up? No one will sit with me. All the other scientists have ostracized me. I sympathize with the Jewish refuseniks. They are scientists who have been ostracized by the Soviet establishment because they applied to emigrate to Israel, were turned down by the Soviet authorities, and now they have no jobs and no place to turn. They are alone like me.[27]

A man whose parents' pride at his scientific achievements was once given full-voice on the front page of Pravda now, three decades later, eating lunch alone.

Graham's encounter with Lysenko occurred around the same time that Medvedev's manuscript was published in the United States. Medvedev was promptly arrested, sent to a mental hospital and diagnosed with schizophrenia, due to his inability to stick to his chosen field (biochemistry), and insistence on writing about history and public affairs (Lysenko) as well.[28]

Dobzhansky sent a letter to Isadore Michael Lerner about what could be done. He added a post-script asking: "Is Lysenko dead?"[29] He was not, and it even seemed, in some unaccountable way, that Lysenko still wielded a sort of power.

In 1972 Dobzhansky appeared in a film shown to a Congress of Geneticists and Selectionists in Moscow. As the film began, Dobzhansky's face and name on the screen drew a ripple of applause from several hundred people in the audience.

---

[26] Correspondence, Martin McCauley to I.M. Lerner, September 17, 1969. B: L563 Lerner. Medvedev, Z.A.—Materials 1969 #4. The American Philosophical Society.

[27] Loren Graham, *Moscow Stories* (Bloomington, IN: Indiana University Press, 2006), pp. 123–4.

[28] Correspondence, Zhores Medvedev to I.M. Lerner June 2, 1970. B: L563 Lerner. Medvedev, Z.A.—Materials 1970 #6. Correspondence, Dr. J. Bonner to I.M. Lerner, July 3, 1970. B: L563 Lerner. Medvedev, Z.A.—Materials 1970 #7. The American Philosophical Society.; Medvedev recounted the incident in a book written with his brother, *A Question of Madness: Repression by Psychiatry in the Soviet Union* (New York: Vintage Books, 1971).

[29] Correspondence, Th. Dobzhansky to I.M. Lerner, June 1, 1970. B: L563 Lerner. Medvedev, Z.A.—Materials 1970 #6. The American Philosophical Society.

Suddenly the film was shut off because, apparently, the sound system had failed. The film was rescheduled for the next day, but the second time the sound track was so badly garbled that when Dobzhansky spoke it was impossible to understand him. Once Dobzhansky's brief cameo ended the technical problems disappeared and the rest of the film was clear and comprehensible.[30]

Dobzhansky attempted a few times to visit the Soviet Union, but was never allowed. The nearest he ever came was the top of a hill in Finland where he could only catch a glimpse.[31] Dobzhansky remained an 'unperson' for the rest of his life. Several times he had visitors from the Soviet Union in New York. One once showed up with an interpreter, a gesture meant to insult him.[32]

## *A Final Flight of Fancy*

In the summer of 1925, while on a late honeymoon in Uzbekistan with his wife, Dobzhansky discovered a species of beetle hibernating in the Kugart Valley. He published his findings in a study, later reprinted in Germany. Many years afterward he came across a paper in a Russian entomological journal on rare beetles. It is customary to give the name of the place where the specimen was collected, the date and the name of the collector. The article cited the location, the date Dobzhansky had been there, but not his name. "I have no doubt whatsoever that there was only one collector on that date collecting beetles in the Kugart Valley," Dobzhansky later said.[33]

Dobzhansky believed there were two kinds of scientists—conservatives and romantics. A conservative is very careful, does not do half-baked stuff—checks and double checks everything, and will not accept a new idea unless the evidence is positively overwhelming. A romantic, meanwhile, proposes many new ideas, most of which are wrong. However what a romantic gets right is brilliant. But no matter what kind of scientist you are, Dobzhansky said, to spend your life trying to prove someone else wrong would be a nightmare. It would be like "hitching yourself to this enemy of yours, in reverse": "Everything he does is by definition wrong." As for himself, Dobzhansky said he always found satisfaction in the fact that he was accused by communists of being a racist, and by racists of being a communist. It provided "symmetry."[34]

---

[30] Correspondence, Professor Mel Greene to Isadore Michael Lerner, February 24, 1972. B: L 563 Lerner. Medvedev, Z.A.—Materials 1972 #1 (Jan.–Apr.). The American Philosophical Society.; "Breakdown in Moscow Blocks Film of Scientist," *The Washington Post*, February 11, 1972; "U.S. Critic of Lysenkoism, in Film, Stirs Soviet Parley," *The New York Times*, February 12, 1972.

[31] Coe, in *The Evolution of Theodosius Dobzhansky*, ed. Adams, p. 26.

[32] The Reminiscences of Theodosius Dobzhansky, p. 325.

[33] The Reminiscences of Theodosius Dobzhansky, pp. 191–2.

[34] B: D65 pt. 2 Dobzhansky on Dunn. Dobzhansky, Theodosius. Reminiscences, Part II: 476; B: D65 Dobzhansky, Theodosius. Reminiscences, Part I: 350. The American Philosophical Society.

Dobzhansky appointed Dunn to be the trustee of his property in the event that his wife pre-deceased him, but Dunn died first. Dunn could have done more it is true, Dobzhansky said, but added, "he could have done more but should not have done more. The way he lived is the way he should've lived."[35]

February 9, 1975, 7:30 p.m., channel 13 in New York aired a Nova special on the Lysenko affair. It was described in the TV guide as "A dramatization of events that culminated in the 1948 Lenin Academy convention, which, at the heeding of biologist T.D. Lysenko, banned the science of genetics in Russia and consequently set that country's agriculture back 20 years."[36]

Five days later Julian Huxley was dead. He had suffered a stroke a few years before, and had been rarely on his feet the last four months of his life. While he was still alive Huxley was brought daily to the Lower Ponds of Hampstead Heath where he found pleasure in observing the water birds.[37] Huxley rejected his grandfather's agnosticism, and declared himself an atheist. Still, he did have some spiritual feeling. In his work *Science, Religion and Human Nature*, published in 1930, Huxley wrote: "At one extreme fear may predominate and force a black and horrifying religion upon men, as in Congo jungles. At the other, reverence and love may be in the ascendant, and make of religion a thing of illumination and joyful peace."[38]

Dobzhansky became religious later in life, although he never believed in a conventional afterward. He said Heaven, if it existed, would not be a place where all the answers about nature would be revealed in a blinding flash. It would be a perpetual existence where experiments would give unambiguous results.[39] Dobzhansky died 10 months after Huxley.

Lysenko outlived them all. His death was announced in a black-bordered box on *Izvestia*'s back page in late November, 1976.[40] A chronicler of Lysenko, V. Safanov, once exulted—"Recall the books you have read by writers like Jules Verne and H.G. Wells, the flights of fancy of imaginative minds—all—even flights of fancy!—pale into insignificance compared with this! ... This is not fancy, however, but actual reality—and we are all participants in it."[41]

This much, at least, was true.

---

[35] B: D65 pt. 2 Dobzhansky on Dunn. Dobzhansky, Theodosius. Reminiscences, Part II: 464. The American Philosophical Society.

[36] "Television This Week," *New York Times*, February 9, 1975.

[37] J.R. Baker, "Julian Sorell Huxley. 22 June 1887–14 February 1975," *Biographical Memoirs of Fellows of the Royal Society* 22 (1976): 232.

[38] Julian Huxley, *Science, Religion and Human Nature* (London: Watts & Co., 1930).

[39] R.C. Lewontin, "Theodosius Dobzhansky, 1900–1975," *Bioscience* 26, no. 2 (1976): 155.

[40] "Lysenko, Science Overlord Under Stalin, Dead at 78," *New York Times*, November 24, 1976.

[41] Safanov, *Land in Bloom*, p. 295.

# Index




W. deJong-Lambert, *The Cold War Politics of Genetic Research: An Introduction to the Lysenko Affair*, Archimedes 24, DOI 10.1007/978-94-007-2840-0,